DIGITAL TECHNOLOGIES FOR SUSTAINABLE FUTURES

This book critically examines the interplay between digitalization and sustainability. Amid escalating environmental crises, some of which are now irreversible, there is a noticeable commitment within both international and domestic policy agendas to employ digital technologies in pursuit of sustainability goals.

This collection gathers a multitude of voices interrogating the premise that increased digitalization automatically contributes to greater sustainability. By exploring the planetary links underpinning the global digital economy, the book exposes the extractive logics ingrained within digital capitalism and introduces alternatives like digital degrowth and the circular economy as viable, sustainable paths for the digital era. Through a combination of theoretical reflections and detailed contextual analyses from Italy, New Zealand, and the UK—including initiatives in participatory planning and technology co-design—it articulates the dual role of digital technology: its potential to support socio-economic and environmental sustainability, while also generating conflicts and impasses that undermine these very objectives. Offering fresh insights into power disparities, exclusionary tactics, and systemic injustices that digital solutionism fails to address, this volume also serves as a reminder that sustainability extends beyond climate-related issues, underscoring the inseparability of environmental discourse from wider social justice considerations.

Aimed at a diverse readership, this volume will prove valuable for students, researchers, and practitioners across various fields, including Geography, Urban Studies, Sustainability Studies, Environmental Media Studies, Critical AI Studies, Innovation Studies, and the Digital Humanities.

Chiara Certomà is an assistant professor of Political-Economic Geography at the University of Turin, Italy. She also serves as a research fellow at the Institute for

Advanced Studies on Science, Technology, and Society (STS) in Graz, at Ghent University, and at the Earth System Governance Research Network at Utrecht University. Her interests include innovative modes of urban governance and planning in response to environmental challenges and the digital turn.

Fabio Iapaolo is a research fellow at Oxford Brookes University's Centre for AI, Culture, and Society, UK. He holds a PhD in Urban and Regional Development from the Polytechnic University of Turin, Italy, and spent a year with the Critical AI Studies (KIM) group at Karlsruhe University of Arts and Design. His work bridges spatial, political, and computer science perspectives to address topics such as algorithmic inequalities, the materiality of computation, and the politics of automation.

Federico Martellozzo is an associate professor of Economic Geography and GIS at the University of Florence, Italy. After earning his PhD in Political and Economic Geography from the University of Trieste in 2010, he served as a postdoctoral fellow at McGill University's Land Use and Global Environment (LUGE) lab in Canada. His research examines the adverse effects of land development and resource consumption patterns amid global environmental changes.

Routledge Studies in Sustainability

Reimagining Labour for a Sustainable Future
Alison E. Vogelaar and Poulomi Dasgupta

Waste and Discards in the Asia Pacific Region
Social and Cultural Perspectives
Edited by Viktor Pál and Iris Borowy

Digital Innovations for a Circular Plastic Economy in Africa
Edited by Muyiwa Oyinlola and Oluwaseun Kolade

Critical Sustainability Sciences
Intercultural and Emancipatory Perspectives
Edited by Stephan Rist, Patrick Bottazzi and Johanna Jacobi

Transdisciplinary Research, Sustainability, and Social Transformation
Governance and Knowledge Co-Production
Tom Dedeurwaerdere

Digital Technologies for Sustainable Futures
Promises and Pitfalls
Edited by Chiara Certomà, Fabio Iapaolo, and Federico Martellozzo

For more information on this series, please visit: www.routledge.com/Routledge-Studies-in-Sustainability/book-series/RSSTY

DIGITAL TECHNOLOGIES FOR SUSTAINABLE FUTURES

Promises and Pitfalls

Edited by Chiara Certomà, Fabio Iapaolo, and Federico Martellozzo

Designed cover image: © Stephen Cornford - a detail from *Projected Landscape #8*

First published 2025
by Routledge
4 Park Square, Milton Park, Abingdon, Oxon OX14 4RN

and by Routledge
605 Third Avenue, New York, NY 10158

Routledge is an imprint of the Taylor & Francis Group, an informa business

British Library Cataloguing-in-Publication Data
A catalogue record for this book is available from the British Library

ISBN: 978-1-032-57854-5 (hbk)
ISBN: 978-1-032-57851-4 (pbk)
ISBN: 978-1-003-44131-1 (ebk)

DOI: 10.4324/9781003441311

Typeset in Times New Roman
by codeMantra

CONTENTS

ABOUT THE CONTRIBUTORS

Ian Babelon is a research fellow at the Department of Architecture and Built Environment, Northumbria University, Newcastle, UK. He is a trained anthropologist and town planner and also works as a UX researcher. He holds a PhD focusing on digital public participation in urban planning.

Gwendolyn Blue is a professor in the Department of Geography at the University of Calgary. Her research examines the social contexts and power dynamics that inform public debates about issues that involve science and technology, including climate change and genomics. Her current research examines public assessments of emerging bio-digital technologies, with a focus on calls for responsible research and innovation for animal and plant genomics.

Guido Boella is a full professor at the Department of Computer Science and also serves as a vice-rector for relations with industries and industrial innovation at the University of Turin. He has authored more than 300 international publications. His research primarily revolves around artificial intelligence and blockchain technology. He is also a co-founder of the Italian Society on Ethics and AI (SIpEIA).

Igor Calzada is a principal research fellow at WISERD at Cardiff University School of Social Sciences, where he leads the WP New Emerging Citizenship Regimes and the Wales & Basque Country Cooperation programme. He has been awarded the Fulbright Scholar-in-Residence (SIR) for 2022–2023 by the US-UK Fulbright Commission at California State University (USA) and was nominated by Apolitical for the list of 100 Most Influential Academics in Government in 2021. His recent monographs, titled *Emerging Digital Citizenship Regimes: Postpandemic Technopolitical Democracies* (2022) and *Smart City Citizenship* (2021), are published by Emerald and Elsevier respectively. He has been selected as an expert for the 'Digital Rights Governance Expert Group: Advisory Support' by the United Nations (UN-Habitat), Cities' Coalition for Digital Rights (CCDR), Eurocities, and UCLG.

Stephen Cornford is a media artist and writer researching the relationships between technologies and landscapes, between media systems and planetary systems. Critically questioning the environmental impacts of consumer electronics and scientific sensing practices, his work questions the viability of addressing ecological collapse through extractive and economic logics. Stephen is currently senior lecturer of fine art at Winchester School of Art and co-director of the Critical Infrastructures and Image Politics research group. He has exhibited work internationally for 15 years, including at the ZKM Centre for Art & Media, (Karlsruhe), ICC (Tokyo), Haus der Electronische Kunst (Basel), and Finnish Museum of Photography.

Mark Dyer is dean of engineering at the University of Waikato, New Zealand. Previously, he held the McNamara Chair in Construction Innovation at Trinity College Dublin, from 2008 to 2019, where he established TrinityHaus as a centre for research and innovation for people, cities, and infrastructure. His earlier professional experience in the 1980s and 1990s involved the design and construction of major infrastructure projects in Europe, Africa, and Asia.

Anne Marte Gardenier has been a PhD student at Eindhoven University of Technology since 2020. In her research, she investigates how society and its citizens can become more cyber resilient in the digitizing society, specifically in the domain of cybersecurity.

Caitlin Hafferty is a postdoctoral researcher in environmental social science at the Environmental Change Institute, School of Geography and the Environment, University of Oxford. Caitlin is broadly interested in addressing interlinked social, environmental, and economic challenges through interdisciplinary and action-oriented research, focusing on meaningful and inclusive participation in environmental governance processes.

Mignon Hagemeijer has been a PhD student at Radboud University since 2023. She investigates the societal and ethical implications of digital health technologies and self-monitoring.

Mél Hogan is the host of The Data Fix podcast and is the director of the Environmental Media Lab (EML). She is an associate professor in the Department of Film and Media at Queen's University. Her research focuses on data infrastructure, extractive AI, and genomic media – each understood from within the contexts of planetary catastrophe, and collective anxieties about the future. For more information, visit: melhogan.com.

Ruth Machen is an environmental and political geographer whose research focuses on knowledge politics during climate science-policy interaction. She currently holds a research fellowship at Newcastle University to develop her research on digital climate governance, examining how algorithmic ways of thinking – embedded in models, platforms, and interfaces – are shaping the process and outcomes of climate governance.

Luis Martin Sanchez, architect (Politecnico di Torino), holds a PhD in urbanism from the Iuav University of Venice. He is currently a postdoctoral research fellow at the University of Turin (Department of Economics, Social Studies, Applied Mathematics and Statistics) and adjunct professor of urban design at the Architecture School of the Politecnico di Torino.

Jessica McLean is a senior lecturer in the School of Social Sciences at Macquarie University where she teaches smart urbanism, Anthropocene politics, and environmental justice. Her book *Changing Digital Geographies: Technologies, Environments and People* (2020) has contributed to shaping the emerging subdiscipline of digital geographies. Jess was founding co-editor-in-chief of the open access *Digital Geography and Society* journal and is currently associate editor of *Transactions of the Institute of British Geographers*.

Irene Niet joined the Eindhoven University of Technology in 2019 as a PhD student. Her research focuses on the governance of artificial intelligence in the energy transition, with a particular interest in the Dutch electricity system.

Margherita Gori Nocentini holds an MSc in sociology from the University of Florence and is currently a PhD candidate in urban planning, design and policy at the Department of Architecture and Urban Studies (DASTU) of the Politecnico di Milano.

Jiří Pánek is an associate professor of regional and social development at the Department of Development and Environmental Studies, Palacký University Olomouc, Czech Republic, with a background in geography, GIScience, and development studies. His main research lies in using GIS in community participation. He is a former Fulbright-Masaryk scholar at the Centre for Geospatial Analyses at North Carolina State University (2021–2022), and a Ruth Crawford Mitchell Fellow in the Urban Studies programme at the University of Pittsburgh (2017).

Claudio Schifanella is an associate professor at the Department of Computer Science, University of Turin. His research encompasses a wide range of topics, including civic technologies, social computing, distributed ledger technologies, data mining, multidimensional data analysis, knowledge representation, and text mining.

Sy Taffel is a senior lecturer of media studies and co-director of the Political Ecology Research Centre at Massey University, Aotearoa-New Zealand. He is the author of *Digital Media Ecologies* (2019). His research focuses on digital technology and the environment, digital media and society, automation, media and materiality, and digital labour. He has also worked as a filmmaker and photographer, and has been involved with media activist projects, including Indymedia, Climate Camp, and Hacktionlab.

Rinie van Est joined the Rathenau Instituut in 1997, where he coordinates research in the field of energy transition and digital transition. Since 2000, he has been working part-time at the Eindhoven University of Technology, where he currently holds the chair of Technology Assessment and Governance.

Cristina Viano is a PhD student in urban and regional development at the Department of Regional and Urban Studies and a project manager at the Department of Computer Science, both at the University of Turin. Her research focuses on the socio-spatial implications of civic technologies in urban contexts and local experiments with blockchain technology.

Min-Hsien Weng is a software engineering researcher with a growing interest in applying machine learning, particularly natural language processing (NLP) and large language models (LLMs), to diverse challenges. His main research interests span software verification, programming languages, and compilers. His recent postdoctoral research fellowship at the University of Waikato in New Zealand ignited a strong interest in using NLP and LLMs to understand complex urban environmental changes. This emerging focus complements his extensive industrial experience in hardware and software integration gained at the Industrial Technology Research Institute (ITRI) in Taiwan, allowing him to bridge the gap between theoretical research and practical applications.

Shaoqun Wu is a senior lecturer in the Computer Science Department at the University of Waikato, New Zealand. She is the research leader and core developer in the FLAX project (flax.nzdl.org). Her research interests include text mining, particularly in the area of automated language pattern extraction and presentation from large volumes of texts, parallel computing for improving software performance, and computer and education.

ACKNOWLEDGEMENTS

This book was produced as part of the 'Digitally-Enabled Social Innovation in the City: Implications for Urban Spaces, Societies, and Governance' PRIN 2022 project (2022KTEZPX), funded by the European Union – NextGenerationEU (Finanziato dall'Unione Europea – Next Generation EU), under the auspices of the Italian Ministry for Universities and Research. The thoughts and reasoning developed in this volume were tested during a small-scale online colloquium titled 'Digital Sustainability? Potentialities and Pitfalls of Digitally-Supported Ecological Transition in Europe.' This event, held on October 11th, 2022, as part of the European Week of Regions and Cities (#EURegionsWeek), was organized in cooperation with the European Regional Science Association (ERSA) and the Association of European Schools of Planning (AESOP). We would like to extend our gratitude to our speakers, including some of the authors featured in this volume, Igor Calzada, Jessica McLean, and Luis Martin Sanchez, as well as to Samantha Cenere and Alberto Cottica who, while not contributors to the book, enriched our colloquium with their presentations. Our thanks also go to all attendees for their valuable participation.

1

DIGITAL (UN)SUSTAINABILITIES

An introduction

Fabio Iapaolo, Chiara Certomà, and Federico Martellozzo

Digital + sustainability

The COVID-19 pandemic has acted as a catalyst for an unprecedented embrace of digital technology across all spheres of life, both public and private. This digital shift—compelled by the need to maintain social and economic continuity amidst global lockdowns and quarantines—has resulted in daily activities such as work, shopping, learning, and socialising transitioning to online platforms. Initially a temporary adaptation, these changes have, over three years or so, become ingrained into a new normal, with digital platforms increasingly playing a key role in shaping our redefined daily reality. The human and social costs of the pandemic are well-known to everyone. Yet, from an environmental standpoint, it has also provided some unexpected respite, albeit brief. With restrictive measures imposed worldwide that slowed production and reduced mobility, to the surprise of many, in April 2020, the world registered the largest decline in CO_2 emissions since World War II (Le Quéré et al., 2020). Such a remarkable event taught us a twofold lesson: first, that even a modest reduction in human activity can have a substantial impact on environmental dynamics; and second, that further digitalisation may, at least in principle, help achieve long-sought environmental goals that have systematically gone unmet for over 50 years. In this sense, despite all its tragedy, the pandemic has also provided a glimmer of hope for our planet's endangered future.

While it's true that the pandemic has renewed emphasis on the link between digitalisation and sustainability, this isn't exactly a novel claim. Indeed, the connection between these two domains has been advocated for years by international entities, as evidenced by a wealth of reports and authoritative endorsements (Santarius et al., 2023; Muench et al., 2022). However, in light of escalating environmental problems that have either neared or already crossed the threshold

DOI: 10.4324/9781003441311-1

of irreversibility (IPCC, 2022), the imperative to integrate digital sustainability into our ethical, political, and economic frameworks has reached unprecedented urgency. As a result, 'digital sustainability' has evolved from 'an ideal about the future' (Hogan & Blue, this volume, p. 34) to an actionable plan, sustained by initiatives such as the European Union's 'twin transition' strategy (EC, 2022). In a world facing mounting environmental challenges, such as species extinction and the cascading effects of glacier retreat, digital sustainability is now at the forefront of the global environmental and computing agenda for the years ahead. This trend is confirmed by a proliferating number of policy guidelines and economic reports, all of which acknowledge as key the integration of digital technologies and environmental science in addressing planetary challenges that require concerted efforts at all governance levels (see Pan & Zhang, 2021; UNEA, 2019; Casal et al., 2004).

Although digital sustainability defies a single definition, it broadly involves leveraging digital technologies, such as infrastructure management systems and communication tools, to foster sustainable development, aligning with the United Nations' Sustainable Development Goals (see George et al., 2021; Mondejar et al., 2021). This includes a variety of strategies and initiatives, from improving energy management and advocating for paperless operations to reducing e-waste and promoting sustainable urban transport. As often claimed, the potential gains extend beyond environmental impacts to fostering equitable societies, enhancing public health, and increasing transparency and accountability across institutions, governments, and businesses. The World Economic Forum highlights how

> today, digital technologies are being used to measure and track sustainability progress, optimize the use of resources, reduce greenhouse gas emissions, and make possible a more circular economy. But digital technologies also enable innovation and collaboration. Artificial intelligence (AI) in design, additive manufacturing and digital twins are some of the powerful tools enabling the next wave of climate change solutions. Internet of Things–enabled sensors, blockchain-based authentication, data-sharing platforms and gamified apps are examples of technologies that foster collaboration across the value chain and align participants on common metrics and goals.
>
> *(Anderson & Caimi, 2022)*

Paraphrasing Jennifer Gabrys (2014, p. 4), this 'becoming environmental' of digitalisation—interpreted here as the use of computation to technologically model and manage natural and societal processes—suggests that digitalisation is not only essential but may also be sufficient on its own for tackling climate-related challenges. This notion rests on the idea that the complex, often obscure networks governing Earth's systems—from resource generation and usage, consumption patterns, and waste production to the interactions between human activity and natural ecosystems—could be deciphered and managed through the integration of advanced computing and transition research. Such an approach is expected to reveal novel exit strategies from the Anthropocene, provided that environmental data is collected, analysed, disseminated, and used effectively by relevant stakeholders. From this perspective, digital sustainability extends beyond the adoption of greener technologies. It signifies, at a more fundamental level, an epistemic paradigm: an array of methods and practices for producing, validating, and applying knowledge to address climate challenges, digital innovations, biodiversity, and their interwoven dynamics.

Digital sustainability has emerged as a collective objective that unites the interests of governments and corporations alike. Sarah Lenz (2021, p. 190) highlights the mutual interest of these sectors in harnessing digital technologies for sustainability, with both often viewing digitalisation as the ultimate 'problem solver' for climate change. For governments, digitalisation not only provides a tool for implementing sustainable strategies and policies but also acts as a catalyst for economic growth. Corporations, for their part, have strong incentives to incorporate environmental, social, and governance considerations into their business practices (see UN Global Compact, 2004), as this can help mitigate negative impacts and open up new avenues for market expansion and profit growth. In this context, the Information and Communication Technology (ICT) industry stands as a key player in addressing 'seemingly intractable problems' (George et al., 2021), both ecological and otherwise, by providing societies with greener technologies and by integrating more sustainable day-to-day business processes and operations (EC, 2021).

But amid the rising enthusiasm, a counter-current of critical examination is gaining momentum within both the social sciences and the computer science community.[1] Scholars from various fields, including critical geography, urban studies, internet and data studies, and political ecology, are voicing scepticism towards the unconditional embrace of what may be aptly termed a 'digital mandate'[2]—a situation Hafferty and colleagues refer to in this volume as an institutionalised 'digital-by-default' attitude. With a growing awareness of how digitalisation can contribute to environmental harm and exacerbate social inequalities (Crawford & Joler, 2018), the oft-assumed symbiotic relationship between 'digital' and 'sustainability' is being confronted on multiple grounds. While, as some chapters in this volume showcase, there exist indeed instances where digitalisation can advance social and environmental justice, it is undeniable that, as it currently stands, digital sustainability remains largely subsumed by the logics of commodification and financialisation that underpin mainstream environmental policies, such as emissions trading, ecosystem services, and green infrastructure (Smessaert et al., 2020; Sullivan, 2013; Lohmann, 2009; Castree, 2003). Simultaneously, as our reliance on digital technologies intensifies, the pursuit of sustainable solutions can paradoxically contribute to existing environmental issues, such as soil, water, and air pollution, while also giving rise to new ones (Kamiya, 2020; Pickren, 2014). A central issue, as Sy Taffel notes in this collection, is what he terms the 'fantasies of dematerialization' associated with the digital economy, a notion that often obscures its reliance on a vast and finite assortment of materials and planetary resources

(Monserrate, 2022; Gabrys, 2011). As a result, there is a growing, interdisciplinary, and somewhat fragmented scholarly endeavour dedicated to critically examining digital sustainability, with findings often at odds with the prevailing narrative. Jessica McLean (2020a, p. 2) captures this sentiment by stating:

> Digital ecosystems are sometimes positioned as a solution to environmental dilemmas without critical reflection of the environmental costs and benefits of the infrastructure and technologies that produce these systems. Discourses of sustainability with respect to digital technologies include assertions of the benefits of paperless offices and frequently do not shift beyond such positions. At the same time, arguments to globalise digital ecosystems for a unified approach to global environmental crises are emerging that are tied to framings of data as a public good but frequently these tend to operationalise digital solutionism. Rhetoric on the need for greater data sharing and transparency of institution-based knowledge is a part of this push. I argue that, despite these optimistic gestures, it is unlikely that data sharing and open digital ecosystems will significantly recast the conditions of the Anthropocene and that such efforts may even further entrench the conditions of this unwanted epoch.

In other words, there is an increasing concern that the purported synergy between 'digital' and 'sustainability' might ultimately prove to be an oxymoron, owing to manifest incompatibilities. But precisely because "neither a future without digital technologies is conceivable, nor is a future devoid of sustainability discourse" (Lenz, 2021, p. 190), it becomes all the more urgent to further interrogate this connection. Without wholly dismissing the potential benefits of digital sustainability or opposing digital computing in itself, this volume is committed to closely examining the intricate dynamics and potential points of conflict between these two areas.

Digital (un)sustainabilities

Today's global trend towards digitalisation, commonly referred to as the 'digital transition', raises a critical question: How does this shift impact environmental and social sustainability? The urgency of this inquiry becomes even more pronounced when considering the extensive socio-technical changes, both current and anticipated, that the digital transition brings about (Chatzistamoulou, 2023; de Bem Machado et al., 2021). As this transition progresses, driven by its potential to fuel economic and technological innovation, stimulate growth, and enable new forms of social interaction (World Economic Forum, 2023), it demands careful steering by policymakers. Central to this process is the need to align a diverse array of stakeholders—and their competing interests—towards common objectives, all while addressing accompanying side effects, including the widening digital divides, increasing privacy concerns, and potential job displacement (García-Peñalvo, 2023; Vial, 2019). Various international policies and domestic development agendas are in place that seek to facilitate this digital transition. The European Commission, for instance, emphasises on its website the transformative role of digital technology in achieving a climate-neutral Europe by 2050, highlighting the need for Europe to strengthen its digital sovereignty and set its own standards in data, technology, and infrastructure (EC, 2023). Examples that underscore this global commitment to digital transformation include:

- European Union's 'Digital Agenda' and 'Digital Single Market': These initiatives aim to unify the digital market across member states, standardise regulations, and improve digital infrastructure.
- China's 'Made in China 2025': Aimed at accelerating digital transformation in the manufacturing sector, this programme prioritises emerging technologies like artificial intelligence (AI), robotics, and 5G.
- India's 'Digital India' Initiative: Focused on increasing digital literacy, expanding digital infrastructure and e-services, and fostering e-commerce and digital innovation.
- USA's 'National AI Strategy': This strategy focuses on advancing research, development, and adoption of AI across various sectors, including healthcare and national defence.
- Singapore's 'Smart Nation' Initiative: Leveraging digital technology and data analytics, this programme aims to enhance urban living, transportation, and healthcare through collaborative efforts involving government, businesses, and citizens.
- Japan's 'Society 5.0': Envisioning a future society integrated with digital technologies, it aims to address societal challenges and stimulate economic growth.
- Canada's 'Digital Charter': Outlining principles for a secure digital environment, it emphasises privacy protection and leveraging digital technologies for economic growth.
- United Nations' Sustainable Development Goals: These goals recognise the role of digital technologies in tackling climate change and promoting environmental protection, while addressing global issues such as poverty, education, and healthcare.

These policies collectively represent a trend where the search for environmental solutions converges with the push for innovation and increased competitiveness in the global digital economy. While the advantages of digitalisation, such as enhanced efficiency and potential energy savings, underpin the rationale for these policies, its impact on sustainable development paints a complex picture that resists simplistic and one-dimensional assessments. Assessing the long-term effects of human actions on ecosystems is indeed particularly challenging due to the multi-temporalities at which climate change phenomena occur. Adding to this complexity are key factors like energy consumption, land use, labour dynamics, and material transformations, which, although geographically dispersed, are interconnected through extensive networks of global chains. Consequently, evaluating the environmental footprint of digitalisation remains inherently relative, requiring a careful weighing of its gains against its drawbacks as they manifest—often disproportionately—across diverse locations and scales.

The role of the ICT industry in global carbon emissions exemplifies this intricate equation. This includes the continual energy consumption throughout the lifecycle of digital technologies, which are not only energy-intensive during operation but also pose significant environmental challenges post-use, particularly considering the limitations of current recycling and disposal methods (Krumay & Brandtweiner, 2016). To mitigate these concerns, the ICT industry has taken active measures, with major internet companies leading the way by transitioning to renewable energy sources and, more controversially, adopting carbon offsetting strategies (Greenpeace International, 2017). Simultaneously, the academic and research community is contributing through initiatives like the 'ICT for Sustainability' (ICT4S) conference series, exploring strategies from reducing consumption patterns to aiming for ambitious targets like carbon neutrality and resource autonomy (D4S, 2022; Lange & Santarius, 2020). Despite the industry's increasing commitment to sustainability, debates continue regarding the efficacy of these initiatives.

While digital advancements are often praised for their eco-friendliness, such as vehicle sharing and automation, it's crucial to acknowledge that they come with concealed environmental costs. These technologies, reliant on electronic systems, can lead to significant rebound effects (Rahman et al., 2023). Alongside these considerations, Smith et al. (2006) emphasise the often-overlooked human cost of the electronics industry, including occupational health hazards. Data centres exacerbate this scenario by imposing considerable energy and water requirements, which not only worsen existing environmental concerns but also introduce new challenges, such as community disruption due to noise pollution (Jones, 2018; Hogan, 2015). The environmental impact of digital currencies like Bitcoin is also notable, with their energy consumption rivalling that of entire nations (IEA, 2019; Mora et al., 2018). Moreover, the digital shift has intensified the demand for raw materials, particularly rare earth elements, leading to new geographies of raw material appropriation (Piscicelli, 2023; Magrini et al., 2021; Massari & Ruberti, 2013). The extraction of these materials, particularly in the Global South, along with the exploitation of a global underclass of on-demand workers in the AI industry (Gray & Suri, 2019), further highlights the deep social and ethical contradictions inherent in digitalisation.

Upon reflecting on the initial promises of digitalisation, a stark contrast emerges between its envisioned ideals and the current reality. Digitalisation, once hailed as a force for democratisation and sustainability, has increasingly become synonymous with technology monopolisation and the emergence of new power enclaves (Tomalin & Ullmann, 2019). Nowhere is the impact of this monopolisation more evident than in urban areas, where the clash between corporate influence and public interests is most pronounced—a situation that initiatives like those discussed in the chapter by Viano and colleagues aim to counter. The dominance of a few corporations in the ICT sector poses a threat to transparency and accountability in sustainability practices, especially due to their cross-border operations and ability to circumvent regulatory oversight (McLean, 2019; Kalbag, 2017). Another concerning issue is the erosion of democratic discourse in the digital era (Caprotti, 2015; Nielsen, 2006). Digital public participation, originally intended to engage a broader population in sustainability policy, often faces technical and cultural barriers, including digital and language issues that mirror existing power dynamics. Simultaneously, it frequently leads to superficial engagement and impoverished decision-making quality (Platteau, 2008; Mohan, 2001). Furthermore, while digitalisation appears to broaden choices in sustainable lifestyles and technologies, it also tends to shift the responsibility for sustainability to the individual, diverting attention from broader structural considerations (Chopra et al., 2023; Strengers, 2013). Taken together, these challenges represent only a fraction of the broader social and environmental complexities in the digital realm, which remains a contentious space for achieving a sustainable, democratic, and inclusive society (Certomà, 2021). As Ruth Machen (p. 93) points out in this collection, "Sustainability has always been about more than just climate". This underscores the need to view digital sustainability as a multifaceted objective, one that requires pursuit through a lens of social justice (Edwards, 2020; McLean, 2020b).

Given this complex backdrop, we are committed to mapping the field to clarify the ongoing debate and establish a research agenda on digital (un)sustainability. Our interdisciplinary approach aims to synthesise insights from various academic disciplines, thereby expanding the understanding of digital sustainability and offering fresh perspectives to the fast-evolving and fragmented critical discourse on this subject. As we embark on this endeavour, we align with scholars who regard digital technology as significant social force, one that is both influenced by and profoundly impacts societal structures in material and symbolic ways (Bijker et al., 2012; Derksen & Beaulieu, 2011; Williams & Edge, 1996). Drawing upon the latest advancements in critical internet studies (Hunsinger et al., 2019), critical urban theory (Marcuse et al., 2014; Brenner, 2009), media studies, and political ecology, among others, this volume seeks to explore life-denying practices, power disparities, exclusionary tactics, and systemic injustice at both global and local levels. A central focus of our investigation is the role of digital capitalism and its capacity to perpetuate these dynamics, reshape urban metabolic processes, and create socio-environmental conditions conducive to its own continuation.

Themes and contributions

This volume assembles a diverse collection of conceptual, theoretical, and empirical chapters, each offering unique perspectives on the complex interplay between digitalisation and sustainability. A common thread running through them is the confrontation of the prevailing notion that greater digitalisation automatically equates to increased sustainability. Part 1, The Uneven Consequences of Digital Capitalism in Global Society, lays the groundwork through both theoretical and case-based contributions dissecting the foundational theories and beliefs underpinning digital sustainability. In Chapter 2, 'Fantasies of Dematerialization: (Un)Sustainable Growth and Digital Capitalism', Sy Taffel critically examines the widely held but flawed belief that digital technologies somewhat enable new forms of dematerialisation. The chapter introduces 'digital degrowth' as a counter to digital capitalism, advocating values like conviviality, decommodification, and growth limits to reduce inequalities, strengthen communities, and revitalise ecosystems. In Chapter 3, 'Big Cloud Solastalgia', Mél Hogan and Gwendolyn Blue critically assess the sustainability initiatives of major tech companies, dismissing these efforts as predominantly 'greenwashing'. Advocating for more profound changes like industry restructuring and reduced profits, they connect their critique to 'solastalgia'—the emotional distress caused by awareness of irreversible environmental damage. Their argument underscores the importance of acknowledging our planet's and our own finite nature as a basis for more communal and sustainable world-building. Chapter 4, 'Operative Landscapes of Digitisation, Collateral Landscapes of Circularity', by Stephen Cornford, conceptualises two distinct types of landscapes: the 'operative', developed as a variation on Harun Farocki's 'operational image', and the 'collateral', which is generated as a byproduct of extractive industries. Cornford makes a compelling case for incorporating waste into technical metabolisms, advocating for a shift away from automating the extraction of new materials and towards recycling waste to establish a genuinely circular economy. Irene Niet, Mignon Hagemeijer, Anne Marte Gardenier, and Rinie van Est present a nuanced perspective on AI's role in sustainability in Chapter 5, 'Framing the (Un)Sustainability of AI: Environmental, Social, and Democratic Aspects'. They explore AI's potential in supporting evidence-based environmental policies and its adverse effects on ecological and social sustainability. The chapter wraps up by suggesting ways to balance digitisation and sustainability transitions, emphasising the necessity to acknowledge AI's environmental costs and promote democratic governance. Chapter 6, 'Problematising Digital Democracy: The Role of Context in Shaping Digital Participation', by Caitlin Hafferty, Jiří Panek, and Ian Babelon, investigates how local contexts affect digital technology's role in participatory planning and environmental decision-making. It examines various hurdles, including access, inclusion, digital literacy, power dynamics, and digital wellbeing. In Chapter 7, 'Digital Fractures: Sustainability and the Partiality of Climate Policy Simulation Models', Ruth Machen addresses the complex task of integrating climate mitigation simulation models into sustainability policy. Machen points out how these models often neglect broader sustainability concerns beyond carbon management, thus exposing the gap between digital simulations and a holistic approach to environmental sustainability.

Part 2, Twin Transition on the Ground: Local Experimentations with Digital Sustainability, delves into specific local case studies to uncover unintended outcomes, such as rebound effects, in development projects that aim to promote sustainability through digital innovation. This part presents a variety of examples that illustrate how digitisation efforts can, ironically, hinder ecological transition. Chapter 8, 'Share an Idea: AI-Augmented Urban Narrative', by Mark Dyer, Shaoqun Wu, and Min-Hsien Weng, discusses the use of the digital platform 'Urban Narrative' by the Christchurch City Council for community involvement in the post-2011 earthquake redevelopment of the city centre. In doing so, the chapter uncovers various interests, power imbalances, and political impasses in participatory urban planning. Chapter 9, 'Data (Un)Sustainability: Navigating Utopian Resistance While Tracing Emancipatory Datafication Strategies', by Igor Calzada, explores the challenges of data sustainability in the digital age, addressing issues of data extractivism, privacy, ethics, and digital rights influenced by big tech. It advocates for a paradigm shift in data governance, emphasising the importance of equitable and sustainable data practices and digital emancipation through initiatives like blockchain-based data architectures and data co-operatives. Chapter 10, 'Embedding Sustainability in Software Design and Development: Accessible Digital Tools for Local Communities', by Cristina Viano, Guido Boella, and Claudio Schifanella, explores the development of digital technologies like geolocated civic social networks and local blockchains to support local social economies and participatory communities. The authors place a strong emphasis on ensuring the political, economic, and socio-cultural sustainability of these technologies throughout their lifecycle. They advocate for open-source tools that avoid speculative data extraction and opinion manipulation, promoting transparent business models and a shift from individualistic to community-based technology applications. Chapter 11 by Luis Sanchez and Margherita Gori Nocentini, 'European Strategic Autonomy for the Twin Transition: Ambiguities and Contradictions from a Spatial Perspective', examines Europe's ecological and digital transformation strategies, focusing on the European Green Deal and New Generation EU. The chapter analyses the 'European Critical Raw Materials Act' through a spatial lens and discusses its complexities, using an Italian re-mining project as a case study to highlight the contradictions and ambiguities in European twin transition policies.

In her concluding chapter, 'Excavating Digital (Un)sustainabilities', Jessica McLean synthesises key insights from the book, focusing on three themes: the gaps and affordances related to digital participation, the debunking of various claims associated with digital solutionism, and the scale and geographies of digital (un)sustainabilities. In the chapter's coda, McLean takes us on a metaphorical journey through the less visible but significant aspects of the digital world, often hidden by its smooth interface. From the mining of cobalt in Australia to the operations of data centres in urban settings, McLean invites us to contemplate the tangible and environmental realities that underpin our digital existence, encouraging a deeper understanding and a reevaluation of the sustainability, or lack thereof, of our digital practices.

Notes

1 As exemplified, for instance, by the 'Computing within Limits' community (see https://limits.pubpub.org/) and the ACM Conference on Human Factors in Computing Systems (CHI) subcommittee 'Critical and Sustainable Computing'.
2 The term 'digital mandate' is adapted from Halpern and Mitchell's (2023) concept of the 'smartness mandate'.

References

Anderson, J., & Caimi, G. (2022). 3 ways digital technology can be a sustainability game-changer. *World Economic Forum*. (January 19). Available at: https://www.weforum.org/agenda/2022/01/digital-technology-sustainability-strategy/ (Accessed 15 November 2023).

Bijker, W. E., Hughes, T. P., & Pinch, T. (Eds.). (2012). *The Social Construction of Technological Systems, Anniversary Edition. New Directions in the Sociology and History of Technology*. Boston: The MIT Press.

Brenner, N. (2009). What is critical urban theory? *City* 13(2–3), 195–204.

Caprotti, F. (2015). Building the smart city: Moving beyond the critiques. *UGEC Viewpoints*. (March 24). Available at: https://ugecviewpoints.wordpress.com/2015/03/24/building-the-smart-city-moving-beyond-the-critiques-part-1/ (Accessed 14 September 2022).

Casal, C. R., Van Wunnik, C., Sancho, L. D., et al. (Eds.). (2004). *The future impact of ICTs on environmental sustainability*. Technical Report EUR 21384 EN. Institute for Prospective Technological Studies at the European Commission's Joint Research Centre (JRC). Available at: https://www.ucc.co.ug/wp-content/uploads/2017/10/The-Future-impact-of-ICTs-and-Environmental-Sustainability.pdf (Accessed 15 November 2015).

Castree, N. (2003). Commodifying what nature? *Progress in Human Geography* 27(3), 273–297.

Certomà, C. (2021). *Digital Social Innovation. Spatial Imaginaries and Technological Resistances in Urban Governance*. New York: Palgrave Macmillan.

Chatzistamoulou, N. (2023). Is digital transformation the Deus ex Machina towards sustainability transition of the European SMEs? *Ecological Economics* 206, 107739. https://doi.org/10.1016/j.ecolecon.2023.107739.

Chopra, R., Agrawal, A., Sharma, G. D., Kallmuenzer, A., & Vasa, L. (2023). Uncovering the organizational, environmental, and socio-economic sustainability of digitization: Evidence from existing research. *Review of Managerial Science* 18, 685–709.

Crawford, K., & Joler, V. (2018). Anatomy of an AI system: The Amazon Echo as an anatomical map of human labor, data and planetary resources. *AI Now Institute and Share Lab*. Available at: https://anatomyof.ai (Accessed 3 September 2022).

de Bem Machado, A., Secinaro, S., Calandra, D., & Lanzalonga, F. (2022), Knowledge management and digital transformation for Industry 4.0: A structured literature review. *Knowledge Management Research & Practice* 20(2), 320–338.

Derksen, M., & Beaulieu, A. (2011). Social technology. In I. C. Jarvie & J. Zamora-Bonilla (Eds.), *The Handbook of Philosophy of Social Science*. London: Sage Publications, pp. 703–719.

Digitalization for Sustainability (D4S). (2022). *Digital Reset. Redirecting Technologies for the Deep Sustainability Transformation*. Berlin: TU Berlin. https://doi.org/10.14279/depositonce-16187.2D4S.

Dryzek, J. S. (2012). *The Politics of the Earth* (3rd ed.). Oxford: Oxford University Press.

Edwards, D. (2020). Digital Rhetoric on a Damaged Planet: Storying Digital Damage as Inventive Response to the Anthropocene. *Rhetoric Review* 39(1), 59–72.

European Commission (EC). (2023). A Europe fit for the digital age. *Empowering people with a new generation of technologies*. Available at: https://commission.europa.eu/strategy-and-policy/priorities-2019-2024/europe-fit-digital-age_en (Accessed 30 October 2023).

European Commission (EC). (2022). *The twin green & digital transition: How sustainable digital technologies could enable a carbon-neutral EU by 2050* (June 29). Available at: https://joint-research-centre.ec.europa.eu/jrc-news-and-updates/twin-green-digital-transition-how-sustainable-digital-technologies-could-enable-carbon-neutral-eu-2022-06-29_en (Accessed 8 May 2023).

European Commission (EC). (2021). *Survey on the contribution of ICT to the environmental sustainability of actions of EU enterprises*. (October 20). Available at: https://digital-strategy.ec.europa.eu/en/library/survey-contribution-ict-environmental-sustainability-actions-eu-enterprises (Accessed 23 April 2022).

Gabrys, J. (2014). *Program Earth: Environmental Sensing Technology and the Making of a Computational Planet*. Minneapolis and London: University of Minnesota Press.

Gabrys, J. (2011). *Digital Rubbish: A Natural History of Electronics*. Ann Arbor: University of Michigan Press.

García-Peñalvo, F. J. (2023). Avoiding the dark side of digital transformation in teaching. An Institutional Reference Framework for eLearning in Higher Education. *Sustainability*, 13(4). https://doi.org/10.3390/su13042023.

George, G., Merrill, R. K., & Schillebeeckx, S. J. D. (2021). Digital sustainability and entrepreneurship: How digital innovations are helping tackle climate change and sustainable development. *Entrepreneurship Theory and Practice* 45(5), 999–1027.

Gray, M. L., & Suri, S. (2019). *Ghost Work: How to Stop Silicon Valley from Building a New Global Underclass*. Boston: Houghton Mifflin Harcourt.

Greenpeace International. (2017). Clicking Clean. (January 10). Available at: https://www.greenpeace.org/international/publication/6826/clicking-clean-2017/ (Accessed 13 November 2020).

Halpern, O., & Mitchell, R. (2023). *The Smartness Mandate*. Cambridge: The MIT Press.

Hilty, L. M., & Aebischer, B. (Eds.). (2015). *ICT Innovations for Sustainability: Advances in Intelligent Systems and Computing*. Cham: Springer.

Hogan, M. (2015). Data flows and water woes: The Utah Data Center. *Big Data & Society* 2(2). https://doi.org/10.1177/2053951715592429.

Hunsinger, J., Allen, M. M., & Klastrup, L. (Eds.). (2019). *Second International Handbook of Internet Research*. New York: Springer.

International Energy Agency (IEA). (2019). *Bitcoin Energy Use - Mined the GAP*. (July 5). Available at: https://www.iea.org/commentaries/bitcoin-energy-use-mined-the-gap (Accessed 20 April 2021).

International Panel on Climate Change (IPCC). (2022). *Climate Change 2022: Impacts, Adaptation and Vulnerability*. Available at: https://report.ipcc.ch/ar6/wg2/IPCC_AR6_WGII_FullReport.pdf (Accessed 15 October 2023).

Jones, N. (2018). How to stop data centres from gobbling up the world's electricity. *Nature* 561(7722), 163–166.

Kalbag, L. (2017). Planning for accessibility. *A List Apart*. (November 09). Available at: https://alistapart.com/article/planning-for-accessibility/ (Accessed 04 June 2021).

Kamiya, G. (2020). Factcheck: What is the carbon footprint of streaming video on Netflix? *Carbon Brief*. (February 25). Available at: https://www.carbonbrief.org/factcheck-what-is-the-carbon-footprint-of-streaming-video-on-netflix (Accessed 3 November 2023).

Krumay, B., & Brandtweiner, R. (2016). Measuring the environmental impact of ICT hardware. *International Journal of Sustainable Development and Planning* 11(6), 1064–1076.

Lange, S., & Santarius, T. (2020). *Smart Green World? Making Digitalization Work for Sustainability*. Abington: Routledge.

Le Quéré, C., Jackson, R. B., Jones, M. W., et al. (2020). Temporary reduction in daily global CO2 emissions during the COVID-19 forced confinement. *Nature Climate Change* 10, 647–653. https://doi.org/10.1038/s41558-020-0797-x

Lenz, S. (2021). Is digitalization a problem solver or a fire accelerator? Situating digital technologies in sustainability discourses. *Social Science Information* 60(2), 188–208.

Lohmann, L. (2009). Climate as investment. *Development and Change* 40(6), 1063–1083.

Magrini, C., Nicolas, J., Berg, H., Bellini, A., Paolini, E., Vincenti, N., Campadello, L., & Bonoli, A. (2021). Using internet of things and distributed ledger technology for digital circular economy enablement: The case of electronic equipment. *Sustainability* 13(9), 4982. https://doi.org/10.3390/su13094982.

Marcuse, P., Imbroscio, D., Parker, S., & Davier, J. (2014). Critical urban theory versus critical urban studies: A review debate. *International Journal of Urban and Regional Research* 38(5), 1904–1917.

Massari, S., & Ruberti, M. (2013). Rare earth elements as critical raw materials: Focus on international markets and future strategies. *Resources Policy* 38(1), 36–43.

McLean, J. (2020a). Frontier technologies and digital solutions: Digital ecosystems, open data and wishful thinking. *Anthropocenes – Human, Inhuman, Posthuman* 1(1), 4. https://doi.org/10.16997/ahip.18.

McLean, J. (2020b). *Changing Digital Geographies: Technologies, Environments and People*. New York: Palgrave MacMillan.

McLean, J. (2019). For a greener future, we must accept there's nothing inherently sustainable about going digital. *The Conversation*. (December 17). Available at: https://phys.org/news/2019-12-greener-future-inherently-sustainable-digital.html. (Accessed 05 June 2022).

Mohan, G. (2001). Beyond participation: Strategy for deeper empowerment. In B. Cooke & U. Kothari (Eds.), *Participation: The New Tyranny?* London: Zed Books, pp. 16–35.

Mondejar, M. E., Avtar, R., Diaz, H. L. B., et al. (2021). Digitalization to achieve sustainable development goals: Steps towards a smart green planet. *Science of the Total Environment*. https://doi.org/10.1016/j.scitotenv.2021.148539.

Monserrate, S. G. (2022). The cloud is material: On the environmental impacts of computation and data storage. *MIT Case Studies in Social and Ethical Responsibilities of Computing*, Winter 2022. https://doi.org/10.21428/2c646de5.031d4553.

Mora, C., Rollins, R. L., Taladay, K., et al. (2018). Bitcoin emissions alone could push global warming above 2°C. *Nature Clim Change* 8, 931–933. https://doi.org/10.1038/s41558-018-0321-8.

Muench, S., Stoermer, E., Jensen, K., Asikainen, T., Salvi, M., & Scapolo, F. (2022). Towards a green & digital future. *JRC Publications Repository*. Available at: https://publications.jrc.ec.europa.eu/repository/handle/JRC129319 (Accessed 17 November 2022).

Nielsen, J. (2006). The 90-9-1 rule for participation inequality in social media and online communities. *Nielsen Norman Group*. (October 8). Available at: https://www.nngroup.com/articles/participation-inequality/ (Accessed 12 August 2022).

Pan, S. L., & Zhang, S. (2021). From fighting COVID-19 pandemic to tackling sustainable development goals: An opportunity for responsible information systems research. *International Journal of Information Management* 55. https://doi.org/10.1016/j.ijinfomgt.2020.102196.

Pearce, F. (2018). Energy Hogs: Can world's huge data centers be made more efficient? *Yale Environment* 360 [online]. Available at: https://e360.yale.edu/features/energy-hogs-can-huge-data-centers-be-made-more-efficient (Accessed 14 September 2022).

Pickren, G. (2014). Geographies of e-waste: Towards a political ecology approach to e-waste and digital technologies. *Geography Compass* 8(2), 111–124.

Piscicelli, L. (2023). The sustainability impact of a digital circular economy. *Current Opinion in Environmental Sustainability* 61, 101251. https://doi.org/10.1016/j.cosust.2022.101251.

Platteau, J.-P. (2008). The pitfalls of participatory development. In *United Nations' Department of Economic and Social Affairs* (Ed.), Participatory Governance and the Millenium Development Goals (MDGs). New York: United Nations, pp. 127–159.

Rahman, M. M., & Thill, J. C. (2023). Impacts of connected and autonomous vehicles on urban transportation and environment: A comprehensive review. *Sustainable Cities and Society*, 104649.

Santarius, T., Dencik, L., Diez, T., et al. (2023). Digitalization and sustainability: A call for a digital green deal. *Environmental Science & Policy* 147, 11–14. https://doi.org/10.1016/j.envsci.2023.04.020.

Smessaert, J., Missemer, A., & Levrel, H. (2020). The commodification of nature: A review in social sciences. *Ecological Economics* 172, 106624.

Smith, T., Sonnenfeld, D. A., & Pellow, D. N. (Eds.). (2006). *Challenging the Chip: Labor Rights and Environmental Justice in the Global Electronics Industry*. Philadelphia: Temple University Press.

Strengers, Y. (2013). *Smart Energy Technologies in Everyday Life: Smart Utopia?* New York: Palgrave Macmillan.

Sullivan, S. (2013). Banking nature? The spectacular financialisation of environmental conservation. *Antipode* 45(1), 198–217.

Tomalin, M., & Ullmann, S. (2019). AI could be a force for good – but we're currently heading for a darker future. *The Conversation*. (14 October). Available at: https://theconversation.com/ai-could-be-a-force-for-good-but-were-currently-heading-for-a-darker-future-124941 (Accessed 21 August 2021).

United Nations Environment Assembly (UNEA). (2019). *The Case for a Digital Ecosystem for the Environment: Bringing Together Data, Algorithms and Insights for Sustainable Development*. Available at: https://un-spbf.org/wp-content/uploads/2019/03/Digital-Ecosystem-final-2.pdf (Accessed 15 November 2023).

United Nations Global Compact. (2004). *Who Cares Wins Connecting Financial Markets to a Changing World*. Available at: https://www.unepfi.org/fileadmin/events/2004/stocks/who_cares_wins_global_compact_2004.pdf (Accessed 23 January 2024).

Vial, G. (2019). Understanding digital transformation: A review and a research agenda. *The Journal of Strategic Information Systems* 28(2), 118–144.

Williams, R., & Edge, D. (1996). The social shaping of technology. *Research Policy* 25(6), 865–899.

World Economic Forum. (2023). *Digital Transition Framework: An Action Plan for Public-Private Collaboration*. Available at: https://www3.weforum.org/docs/WEF_Digital_Transition_Framework_2023.pdf (Accessed 15 November 2023).

PART 1

The uneven consequences of digital capitalism in global society

2

FANTASIES OF DEMATERIALIZATION

(un)sustainable growth and digital capitalism

Sy Taffel

Introduction

As the editors make clear in the introduction, digital technologies are widely purported to resolve a range of contemporary ecological crises: climate change, the sixth mass extinction of life on earth (the biodiversity crisis), plastic and chemical pollution, disruptions to the nitrogen and phosphorus cycles, and land-use change, the crises associated with breaching planetary boundaries (Steffen et al. 2015) and the Anthropocene (Zalasiewicz et al. 2017). This perspective on digital sustainability depends upon the belief that these technologies exemplify a dematerialized, post-industrial logic. This assumption is deeply flawed, constituting what I term 'the fantasy of dematerialization'. This chapter argues that this fantasy plays a key role in enabling a misguided discourse which contends that digital capitalism facilitates perpetual, compound economic growth on a materially finite planet. The fantasy of dematerialization diverts attention from meaningfully addressing ecological crises, which requires reducing global material and energy usage. In unpacking why this fantasy has become commonplace, I point to five interconnected socio-technical factors, wirelessness, miniaturization, the invisibility of digital infrastructures, and the multi-scalar temporalities of digital technologies.

If dematerialization, and the accompanying discourse of infinite green growth on a materially finite planet, is a fantasy, then digital capitalism is fundamentally incompatible with sustainability and social justice. Addressing this contradiction, the chapter advocates supplanting digital capitalism with a degrowth or post-growth approach that challenges assumptions that infinite growth is possible or desirable, while advocating for an equitable redistribution of wealth and resources. I outline conviviality, decommodification and limits as useful guiding principles for digital degrowth. Degrowth enables a radical re-envisioning of digital technologies that

DOI: 10.4324/9781003441311-3

addresses inequalities, connects communities, and revitalizes ecosystems, whereas their current orientation predominantly centres growing corporate profits while promoting inequality, precarity, and competitive individualism. Before turning to the fantasy of dematerialization and digital degrowth though, I begin by briefly outlining the contours of digital capitalism and green growth.

Digital capitalism and dematerialized (green) growth

'Digital capitalism' signals a shift towards a specific form of capitalism where networked digital technologies form 'the central production and control apparatus of an increasingly supranational market system' (Schiller 1999: xiv). Digital capitalism is synonymous with the processes and practices of neoliberalization (Dean 2009, Beer 2016, Williams and Gilbert 2022), which as economic geographer David Harvey outlines 'requires technologies of information creation and capacities to accumulate, store transfer, analyse, and use massive databases to guide decisions in the global marketplace' (Harvey 2005: 3). Digital technologies are integral to the logistical infrastructures required for just-in-time production, and the spatio-temporal acceleration of the circuits of production and consumption associated with globalized consumer capitalism (Berardi 2009, Crary 2013). More recently, digital capitalism is associated with the discourses of platform (Srnicek 2016), surveillance (Zuboff 2019), and communicative (Dean 2009) capitalism, which foreground specific elements of contemporary digital technology's hegemonic role within contemporary capitalism.

Digital capitalism acknowledges empirical findings indicating multiple breaches to planetary boundaries, but its advocates adopt an ecomodernist position (Asafu-Adjaye et al. 2015, Latour 2016), contending that the boundless creativity of technological innovation will resolve these issues. Far from comprising a fundamental contradiction of capitalist accumulation, as is claimed by eco-Marxists (Moore 2015, Foster 2022), the Anthropocene is understood as a business opportunity for 'green growth'. A societal transition towards renewable energy and nuclear power, alongside the retention of fossil fuels whose greenhouse gas (GHG) emissions are offset by technologies, including carbon capture and storage and direct air capture, is understood as an opportunity for shrewd businesses to flourish. This will allegedly usher in the 'good Anthropocene' (Hamilton 2016), whereby rational human stewardship affords ecological stability alongside enduring economic growth, which will be successfully decoupled from presently observed correlations with energy use and material footprint (Wiedmann et al. 2015, Calatayud and Mohkam 2018).

The model of technological solutionism (Morozov 2014, Taffel 2018) underpinning digital capitalism has a superficial appeal that masks the enormous risks associated with continuing attempts to pursue ongoing compound (exponential) increases in gross domestic product (GDP). This year-on-year accumulation of surplus value is integral to capitalism, but has led to ecological overshoot. Digital capitalism's appeal is rooted in claims that 'green growth' and technological innovation lead to a smart, digitally mediated future, where the rapacious levels of consumption associated with contemporary billionaires can allegedly become a generalized state of existence for humanity. Such claims thoroughgoingly internalize the bourgeois ideology which equates wealth with private commodity ownership. However, for immense numbers of people for whom neoliberal austerity has produced financial insecurity, precarity, and material hardship, the promise of a future containing dramatically increased material prosperity alongside successful resolutions of ecological crises is undoubtedly attractive (Huber 2022).

The issue is that this position places an enormous degree of faith in technological progress that lacks empirical evidence. While some solutionist panaceas, such as Elon Musk's vision of colonizing Mars (Kulwin 2016), are so far-fetched that they should simply be dismissed outright, other forms of solutionism are more plausible, but are likely to inflict additional environmental and social harms. Nonetheless, a lack of evidence does not guarantee failure; neo-Malthusian arguments that fixed natural barriers cannot be overcome have typically failed to account for capitalism's dynamism, its struggle to overcome any apparent limits to an ever-increasing rate of accumulation. As Marx (1973: 334/335) argued, 'capital is the endless and limitless drive to go beyond its limiting barrier. Every boundary is and has to be a barrier for it'. This strategy of circumnavigating apparent boundaries is discussed by ecosocialists as a metabolic shift (Moore 2017), whereby the consequences of the metabolic rift between capitalism and ecology are temporarily addressed by displacing negative socio-ecological effects.

Metabolic shifts occur by displacing harms across time and space, or by adopting new technologies (Saito 2023). An insightful example of the latter is the 'resolution' of peak oil. During the late-twentieth-century peak oil, denoted concern that as discoveries of crude oil dwindled, potential shortages would hamper social development (Bridge 2010). Rather than presenting a hard limit, however, this led to a technological shift towards 'unconventional oil' derived from sources such as tar sands or using techniques such as hydraulic fracturing (commonly known as fracking). These novel methods for obtaining oil are more energy intensive than conventional methods, meaning they are worse in terms of contributing to climate change, and they create novel health and environmental threats, ranging from contaminating drinking water – including the spectacular images of flammable drinking water evidenced in the film *Gasland* (2010) – to mercury and arsenic pollution resulting from tar sands extraction (Timoney and Lee 2009). While the 'crisis' of peak oil was averted, significant new harms were inflicted by this resolution.

Homologously, the discourse of green growth suggests that an energy transition (partially) moving away from fossil fuels will 'resolve' the climate crisis. However, this ignores the harms associated with massively expanding extractive activities to produce sufficient energy to not only replace the 85% of global energy generated from fossil fuels in 2021 (Ritchie and Roser 2021) but also additionally facilitate ongoing compound growth of that figure. This would dramatically increase the overshoot of planetary boundaries involving biodiversity loss, land-system change, and chemical pollution alongside numerous localized harms associated with mining. Further, several recent analyses indicate critical shortages of specific metals, such as cobalt, nickel, and copper needed for a global roll-out of renewable technologies (Azevedo et al. 2018, Michaux 2021). While doubts surrounding the viability of the requisite expansion of short-term extraction are somewhat valid, the neo-Malthusian framing largely neglects the possibility of shifting resource requirements, either technologically, such as using lithium-iron-phosphate batteries to reduce the need for nickel and cobalt (Zeng et al. 2022), or spatio-temporally, by employing deep-sea mining to extract cobalt from polymetallic nodules from the oceanic floor (Sharma 2017). The latter process displaces harms to remote and poorly understood ecosystems, where substantial harms may cascade across oceanic ecologies (Washburn et al. 2019, Bedford, McGillivray, and Walters 2020). This denotes a spatial shift towards harming the oceans and a temporal shift towards harming the future. Spatio-temporal displacements do not resolve harms they reallocate them, enabling the short-term continuation of globalized capitalism, whose ongoing demands for increased extraction, accumulation, energy, and materials are fundamentally incompatible with a sustainable future.

While green growth alludes to decoupling GDP growth – growth in quantitative exchange value – from material use and GHG emissions, there is no empirical evidence this decoupling can take place at the pace required to meet international commitments surrounding climate change in terms of GHG, or that decoupling can take place at all in the case of material footprint (Bringezu 2015, Calatayud and Mohkam 2018, Hickel and Kallis 2019). However, digitally enabled 'green growth' remains the dominant discourse for addressing global ecological crises. Given the manifold issues associated with technological solutionism, a key fantasy underpinning this discourse surrounds dematerialization, the idea that digital technologies enable infinite economic growth without corresponding increases in materials or energy.

The fantasy of dematerialization

Digital technologies have long been associated with linguistic tropes suggesting dematerialization: 'virtual reality' and 'virtual communities' (Rheingold 1991, 1993), 'immaterial labour' (Lazzarato 1996) and more recent concepts such as 'cloud computing' (Amoore 2020). These metaphors suggest digital technologies have a fundamentally different relationship with materiality than industrial technologies, fuelling the erroneous notion that digital technologies facilitate decoupling. These metaphors both support and represent the dematerialized view of digital technologies that are central to discourses of green growth, technological solutionism, and digital capitalism. However, these metaphors are not just a linguistic sleight-of-hand. While they play an ideological role in supporting growth-focussed strategies for addressing ecological crises, they are also rooted in certain properties and affordances of digital technologies.

The first of these factors is wirelessness (Mackenzie 2010). While data infrastructures overwhelmingly rely upon transmission through undersea and underground fibre-optic cables (Starosielski 2015), the last step of this communication process involves smartphones, tablets, laptops, internet of things (IoT), and wearable devices employing wireless 4/5G cellular or Wi-Fi connections. Consequently, contemporary digital technologies are often experienced as being wireless. Wirelessness feeds the fantasy of dematerialization because there appears to be no tangible connection between devices and networked infrastructures. Digital technologies appear to work by magic; indeed, this is precisely why advertisements for digital technologies such as Apple's AirPods equate wirelessness with both freedom and magic (Taffel 2023).

Wirelessness alone, however, is insufficient to conjure the fantasy of dematerialization. Earlier wireless media technologies, such as radio and television, similarly modulated electromagnetic waves and wirelessly transmit signals, but visible antennae on houses, cars, and portable receivers provided prominent visual reminders of the process of signal transmission and reception. Contemporary digital technologies typically obscure this process. Antennae within phones, tablets, and laptops are concealed within black-boxed devices, while cell towers are routinely camouflaged or disguised, assisting in their functional invisibility (Parks 2009, Farman 2014). Here we see the convergence of two additional phenomena that propagate the fantasy of dematerialization, the miniaturization of microelectronic technologies and the invisibility of digital infrastructures.

Miniaturization has long been integral to digital technologies. It is why the computer in your pocket is more computationally powerful than the room-sized mainframe computers of the 1970s. Whereas the Intel 8086 microprocessor used in the original 1978 IBM PC contained 29,000 transistors with 3,200 nanometres long silicon channels (Mueller 2006), contemporary flagship smartphone chips such as the Qualcomm Snapdragon 8 Gen 2 or Apple A16 Bionic contain approximately 16 billion four nanometres long transistors (Cotta 2023).[1] A nanometre is one billionth of a meter. For reference, a single human hair is approximately 100,000 nanometres wide. The microscopic spatial scales at which digital technologies operate are astounding feats of engineering, but this deeply inhuman size is frequently conflated with dematerialization.

These technologies light-weight, portability, and computational power present an aura of dematerialization because they seemingly defy physics. However, this largely arises due to our engagement with them as commodities, as discrete objects, rather than their enormously complex and varied processes of production. Smartphones typically contain around 70 elements, many of which are only found in commercially viable deposits in a handful of locations on Earth. Despite weighting approximately 150 grams, smartphones require around 70 kg of raw materials, around 450 times the weight of the final device (IEEE 2013). The energy and material intensive industrial processes that produce ultra-purified materials for high-performance contemporary microelectronics, such as nine-nines silicon (99.999999999% pure) or electrolytically refined copper (>99.95% pure) entail that roughly 80% of a smartphone's lifetime GHG emissions – between 50 and 100 kg of CO_2 emission for a flagship iPhone (Apple 2022) – result from production.

Digital technologies do not function as isolated objects, they rely upon enormous infrastructures. Infrastructures are 'by definition invisible, part of the background' (Leigh-Star 1999, 380) that enables the functioning of everyday life. They usually only become visible when they fail to function (Jackson 2014; Vinsel and Russell 2020). While we point to smartphones as 'technology', the overwhelming majority of their functionality depends upon gargantuan infrastructural assemblages. Digital infrastructures encompass networks of hardware, including datacentres that store zettabytes of information and perform computationally hyper-intensive activities such as training and inference for deep-learning-based AI systems, millions of kilometres of optical fibres through which data is transmitted, cable landing stations, internet exchange points, 4/5G base stations and antennae, and Wi-Fi routers and repeaters required for wireless networks. Digital infrastructure additionally includes a software layer including firmware, operating systems, applications, programming languages, APIs etc., and protocols and standards including the Internet Protocol, Transmission Communication Protocol, Hypertext Transfer Protocol, Wi-Fi and Bluetooth standards, among many others. Alongside the vast technological assemblage, which this inexhaustive list only begins to document, we should recognize the enormous volumes of labour and knowledge required to constitute and maintain this planetary network. We mistakenly approach digital devices as 'our' individual, isolated commodity-artefacts rather than situating them within infrastructural ecologies. Exploring the flows of energy and matter associated with digital assemblages foregrounds relational processes rather than discrete objects, and reveals intensive flows of fossil fuels, electricity, and matter that stand in glaring contrast to the fantasy of dematerialization.

A range of inhuman spatial scales homologously permeate the temporalities of digital technologies. The capacity of microelectronics to perform astonishing volumes of calculation in imperceptibly brief periods, exemplified by Apple's A16 chip's reported 17 trillion operations per second, further entrenching the sense of magic and dematerialization, especially when combined with miniaturization. At the other end of the temporal scales at which digital technologies operate lie the environmental harms associated with GHG emissions and toxic waste. Digital waste is only partially captured by the 54 million tonnes of e-waste discarded in 2020 (Forti et al. 2020). This does not include waste associated with producing technologies and infrastructures, which as we have seen, in many cases exceeds 99% of the mass of the final 'product'. Elements of these waste streams, including PFAS (Zhao et al. 2023) – commonly known as 'forever chemicals' as they take up to a thousand years to degrade – and plastics (Farrelly et al. 2021) persist and enact harms over durations that greatly exceed embodied human temporalities.

The interplay of these factors perpetuates the fantasy of dematerialization. We employ linguistic tropes that segregate digital technologies from material reality. Wirelessness means that our digital devices appear untethered from the earth, reinforcing the discursive obfuscation of 'the cloud'. Miniaturization, in association with the micro-temporal durations of information processing and communication, entails that digital technologies operate on spatio-temporal scales beyond human perception. The invisibility of infrastructure, which is aided by wirelessness, reinforces the aberrant notion that digital technologies are discrete commodities. Commodity thinking, infrastructural invisibility and spatial remoteness associated with globalized supply-chains obscure the enormous apparatus of production required for digital technologies and the ecological harms this apparatus inflicts. Finally, the slowness of environmental harms associated with climate change, forever chemicals and plastics are not easily perceived, and often occur in places far removed from urban centres of technological consumption.

The problem of how to perpetuate compound economic growth in a materially and energetically sustainable society therefore cannot be technologically fixed by using ever-increasing volumes of digital technology. Consequently, looking beyond the fantasy of dematerialization means jettisoning green growth-based approaches. This requires a serious engagement with degrowth and post-growth perspectives.

Digital degrowth

Put succinctly, degrowth advocates for a planned reduction of material and energy use in order to prevent ecological overshoot of planetary boundaries, while rapidly reducing material inequality between and within nation states, and enhancing human wellbeing (Kallis 2018, Hickel 2021, Schmelzer, Vetter, and Vansintjan 2022).

The first clause of this definition addresses ecological crises, contending it is highly improbable that green growth and technological solutionism will resolve ecological overshoot. Since 1990, when climate change was first widely recognized, GHG emissions have risen from 38 to 55 billion tons (Ritchie, Roser, and Rosado 2020). While a 50% emissions reduction by 2030 is now emphasized as ecologically necessary, emissions have steadily grown, with the only fleeting reductions corresponding to global recessions.[2] While GHG emissions are decoupled from energy use when replacing fossil fuels with renewable energy, there is no evidence that a transition can take place rapidly enough within the context of growing aggregate energy demand (Hickel and Kallis 2019). This requires year-on-year reductions exceeding those experienced during 2020, when much of society was confined during lockdowns. Further, climate change is only one planetary boundary presently being overshot. Material use is a key driver of biodiversity loss (IPBES 2019) and currently there exists no evidence that it is possible to decouple material footprint from GDP growth. 'Renewable' energy requires non-renewable materials, and during the last three decades, the rate of material footprint growth has exceeded that of GDP (Hickel 2020, Wiedenhofer et al. 2020).

The second defining feature of degrowth emphasizes equity and wellbeing, denoting a further shift away from the capitalist mode of production. While critics castigate degrowth as continuing neoliberal austerity (Phillips 2015, Huber 2022), such claims apparently fail to grasp that neoliberal austerity has overseen the growth of material and energy use, accompanied by widening material inequality as the overwhelming majority of wealth accumulates with the global rich (Oxfam 2020). Degrowth, by contrast, aims to reduce material and energy usage while radically curtailing inequality. Notably, this does not focus on GDP, a quantitative measure of monetary exchange, which degrowthers argue provides a poor proxy for the wealth of a society once basic needs are met (Hickel 2020, Jackson 2021). Degrowth argues that forms of radical abundance which are inimical to capitalism, can flourish if societies focus on equity and ecological balance (Hickel 2019, Soper 2020). This involves social policies, including a universal basic income, universal public services, job guarantees, and free education, that are designed to mitigate the social impact of curtailing capitalist industries that actively cause harm, such as the arms trade, or those predominantly geared towards generating exchange and surplus value rather than genuinely meeting human needs, such as much of the financial sector. While degrowth at times emphasizes decentralization, cooperatives, and direct democracy, it should be noted that the kind of centralized planning involved in these policies requires an active role for the state, aligning degrowth with ecosocialism (Kallis 2019a, D'Alisa and Kallis 2020, Albert 2023).

With regard to technology, degrowth draws inspiration from Ivan Illich's (1973) distinction between 'convivial' tools that promote autonomy, equity, and justice, and industrial tools, which compel humans to adapt to centralized, disempowering and dehumanizing logics, and require specialized knowledge and hierarchies to operate. Conviviality cannot be easily reconciled with digital technologies, which as we have seen, require global supply chains and complex industrial processes, and produce vast amounts of (often toxic) waste and GHG throughout their lifecycle. Here it is worth recalling Illich's (1973, 37) argument that

> It is a mistake to believe that all large tools and all centralized production would have to be excluded from a convivial society. It would equally be a mistake to demand that for the sake of conviviality the distribution of industrial goods and services be reduced to the minimum consistent with survival.

Illich's rejection of purity demarcates a pathway for tools that are in certain ways inimical to conviviality to exist within a convivial society. Orienting digital technology towards conviviality begins by recognizing the multitude of harms associated with current digital systems, from mining and manufacturing, through to the design of attention-grabbing, algorithmically curated and advertising-saturated social media platforms and games, and designing strategies to enhance conviviality and curtail harm wherever possible.

A second key strategy for degrowth and digital technology surrounds decommodification. This denotes a concerted effort to remove goods and services from the privatized domain of the market, and instead expand models of common, public, and other forms of non-profit based collective ownership. Under capitalism, the inverse situation dominates. Over time, more and more of everyday life and the biosphere are rendered into quantifiable and exchangeable monetized commodities. Whereas at the dawn of capitalism this involved the enclosure of commons and theft of indigenous lands, knowledge and taonga (treasures), digital capitalism has seen the commodification of communication on corporate social media platforms, and various forms of dataveillance, including locational and biometric data, forming 'data colonialism' (Couldry and Mejias 2019). Longstanding analyses emphasize that capitalism always requires an 'outside' to appropriate wealth from, and to externalize harms towards (Luxemburg 1951). With few spaces left for capitalism to colonize – aside from the deep ocean floor or other planets – data has become a key new frontier for commodification.

Degrowth contends that decommodification of many facets of digital technology, replacing for-profit, corporate services with not-for-profit alternatives that construct commons and public goods, creates preferable environmental and social outcomes. At the levels of software, protocols and applications, this involves the utilization and expansion of free and open-source software and protocols. A good example is Activity Pub, the open social media protocol whose best-known application is Mastodon, a decentralized alternative to Twitter. Whereas Twitter is a for-profit, privately owned service whose business model is predicated upon dataveillance and algorithmically targeted advertising, Mastodon is a collective of federated servers with no advertising. Consequently, it does not require much of the materially and energetically intensive infrastructure necessary for Twitter's profit-driven system (Laser et al. 2022). Decommodification at the level of infrastructure could involve the delivery of broadband internet as a universal public service, rather than an individualized commodity. Essentially this would involve treating Internet access like a publicly funded and maintained transportation network, entailing that the benefits of high-speed internet access were universally available. While decommodifying the production and supply chain of digital technologies is a complex task, initiatives such as Chile's plans to nationalize its lithium industry suggest tentative steps towards forms of extraction geared towards utilizing mineral wealth to enhance national economies and incorporating environmental protections (Villegas and Scheyder 2023).

Alongside conviviality and decommodification, voluntarily embracing limits is an important marker of degrowth. Whereas neo-Malthusian approaches contend that insurmountable external limits necessitate change, degrowth argues that limiting certain things should be welcomed, not as an imposition but because consciously choosing to adopt limits enables other phenomena to flourish (Kallis 2019b). A useful analogy is choosing to limit vehicular speeds outside schools, not because cars cannot drive at speeds guaranteed to be lethal to inattentive schoolchildren but to reduce the likelihood of fatal accidents. Degrowth analogously argues for limiting individual wealth, so that there can be well-funded universal public services, and limiting GHG emissions and material use, so that ecosystems can thrive in the future.

In terms of limiting digital activities, massively curtailing advertising is a key demand of degrowth (Hickel 2020, Taffel 2023). Enormous volumes of labour, energy, and matter are currently dedicated to technologically and psychologically sophisticated attempts to not just normalize, but actively grow an ecocidal circuit of production/consumption. Limiting this activity so that ecological overshoot can be remediated is urgently required. Similarly, limiting excessively data-hungry activities such as 4K/8K streaming video and cryptocurrency mining are relatively straightforward ways to massively reduce the material and energy footprints of digital technologies, while doing little to reduce the use-values communities derive from technology. Additionally, limiting practices such as planned obsolescence would slow the rapacious pace of upgrade culture, reducing the energy and material requirements needed for maintaining digital systems. The point is to embrace limits to reduce the serious, long-lasting ecological and social harms currently associated with digital technology, and to prevent those harms from continuing to grow, as per their current trajectory.

This relational formulation of limits should not be understood in simplistic terms of 'less' or 'more', as has been frequently seen in debates surrounding degrowth and ecomodernism (Hickel 2020, Robbins 2020, Huber 2022). The purpose of voluntarily accepting limits is not to have less of everything, but to have less of undesirable things – advertising, dataveillance, bitcoin, irreparable devices, PFAS, etc. – and more desirable ones – universal broadband access, time with loved ones, pursuing hobbies, and thriving ecological systems. This should be understood as a form of radical digital abundance, whereby abundance substantively departs what we have today, an abundance of spam, phishing, viruses, surveillance, disinformation, and advertising, but instead where digital abundance can be re-envisioned to related to forms of mutual aid, care, and community-oriented communication.

Conclusion

The fantasy of dematerialization plays a pivotal role underpinning the discourse that infinite, compound, green growth is possible, even while current levels of GHG and material use have overshot planetary boundaries, manifesting in the now daily news of droughts, floods, fires, and extinctions. This fantasy is predicated on the confluence of wirelessness, miniaturization, invisible infrastructures, and the more-than-human temporalities of digital technologies. Any just response to contemporary ecological crises requires moving beyond this fantasy, towards a post-growth or degrowth perspective where finitude is acknowledged and accepted.

Digital technologies are ecologically costly and harmful artefacts. The levels of material and energy intensity required for high-performance computing are in some ways inimical to conviviality and sustainability. However, the popularity and utility of these technologies, as well as their necessity for undertaking activities such as balancing electricity supply/demand across renewable-powered grids, means that any democratic model of degrowth must move beyond moral puritanism in accommodating these technologies, while systematically working to reduce the multitude of harms that accompany them. Conviviality, decommodification, and limits are useful frameworks for orienting digital technologies towards post-growth futures, ones that centre ecological balance and sustainability alongside social justice, rather than capitalist accumulation, corporate profits, and the impossibility of infinite, compound growth on a finite planet.

Notes

1 This measurement has shifted from the smallest element in the transistor towards approximating 2D equivalent sizes for stacked 3D designs, leading to manufacturer's specifications being empirically duplicitous marketing ploys. For example, the Intel 14nm and TSMC 7nm processes have almost identical transistor sizes (Tyson 2020).
2 Resulting from the 1997 Asian financial crisis, 2007/8 global financial crisis and 2020 COVID-19 pandemic.

Bibliography

Albert, M.J. (2023) 'Ecosocialism for Realists: Transitions, Trade-Offs, and Authoritarian Dangers'. *Capitalism Nature Socialism* 34 (1), 11–30

Amoore, L. (2020) *Cloud Ethics: Algorithms and the Attributes of Ourselves and Others*. Durham and London: Duke University Press

Apple (2022) *IPhone 14 Pro Max Product Environmental Report* [online] available from <https://www.apple.com/nz/environment/pdf/products/iphone/iPhone_14_Pro_Max_PER_Sept2022.pdf>

Asafu-Adjaye, J., Blomquist, L., Brand, S., Brook, B.W., DeFries, R., Ellis, E., Foreman, C., Keith, D., Lewis, M., Lynas, M., Nordhous, T., Pielke, R., and Shellenberger, M. (2015) *An Ecomodernist Manifesto.*available from <https://static1.squarespace.com/static/5515d9f9e4b04d5c3198b7bb/t/552d37bbe4b07a7dd69fcdbb/1429026747046/An+Ecomodernist+Manifesto.pdf> [14 September 2022].

Azevedo, M., Campagnol, N., Hagenbruch, T., Hoffman, K., Lala, A., and Ramsbottom, O. (2018) *Lithium and Cobalt* [online] McKinsey. available from <https://www.mckinsey.com/~/media/mckinsey/industries/metals%20and%20mining/our%20insights/lithium%20and%20cobalt%20a%20tale%20of%20two%20commodities/lithium-and-cobalt-a-tale-of-two-commodities.pdf>

Bedford, L., McGillivray, L., and Walters, R. (2020) 'Ecologically Unequal Exchange, Transnational Mining, and Resistance: A Political Ecology Contribution to Green Criminology'. *Critical Criminology* 28 (3), 481–499

Beer, D. (2016) *Metric Power*. London: Palgrave Macmillan

Berardi, F. (2009) *The Soul at Work: From Alienation to Autonomy*. Los Angeles: Semiotext (e)

Bridge, G. (2010) 'Geographies of Peak Oil: The Other Carbon Problem'. *Geoforum* 41 (4), 523–530

Bringezu, S. (2015) 'Possible Target Corridor for Sustainable Use of Global Material Resources'. *Resources* 4 (1), 25–54

Calatayud, P. and Mohkam, K. (2018) *Material Footprint: An Indicator Reflecting Actual Consumption of Raw Materials* [online] available from https://www.statistiques.developpement-durable.gouv.fr/sites/default/files/2018-10/datalab-essentiel-142-empreinte-matiere-eng-avril2018b.pdf

Castells, M. (2011) *The Rise of the Network Society: The Information Age: Economy, Society, and Culture 2nd Edition.* vol. 1. Oxford: Wiley-Blackwell

Cotta, R. (2023) *Snapdragon 8 Gen 2 vs A16 Bionic - Which Is the Better Chip?* [online] available from <https://www.videogamer.com/tech/smartphone/snapdragon-8-gen-2-vs-a16-bionic/> [25 May 2023]

Couldry, N. and Mejias, U.A. (2019) *The Costs of Connection: How Data Is Colonizing Human Life and Appropriating It for Capitalism.* Stanford, CA: Stanford University Press

Crary, J. (2013) *24/7: Late Capitalism and the Ends of Sleep.* London and New York: Verso

D'Alisa, G., and Kallis, G. (2020) 'Degrowth and the State'. *Ecological Economics* 169, 106486

Dean, J. (2009) *Democracy and Other Neoliberal Fantasies: Communicative Capitalism and Left Politics.* Durham: Duke University Press

Farman, J. (2014) 'The Materiality of Locative Media: On the Invisible Infrastructure of Mobile Networks'. In Herman A., Hadlaw J., Swiss T. (Eds.), *Theories of the Mobile Internet* (pp. 45-59). New York: Routledge

Farrelly, T., Taffel, S., and Shaw, I. (2021) *Plastic Legacies: Pollution, Persistence, and Politics.* Edmonton: Athabasca University Press

Forti, V., Balde, C.P., Kuehr, R., and Bel, G. (2020) *The Global E-Waste Monitor 2020: Quantities, Flows and the Circular Economy Potential*United Nations University/United Nations Institute for Training and Research, International Telecommunication Union, and International Solid Waste Association available from <https://ewastemonitor.info/wp-content/uploads/2020/11/GEM_2020_def_july1_low.pdf [12 May 2023]>

Foster, J.B. (2022) *Capitalism in the Anthropocene: Ecological Ruin or Ecological Revolution.* New York: NYU Press

Gasland: Can you light your water on fire? (2010) Josh Fox. [DVD] New York: Docurama Films

Hamilton, C. (2016) 'The Theodicy of the "Good Anthropocene"'. *Environmental Humanities* 7 (1), 233–238

Harvey, D. (2005) *A Brief History of Neoliberalism.* Oxford: Oxford University Press

Hickel, J. (2021) 'What Does Degrowth Mean? A Few Points of Clarification'. *Globalizations* 18 (7), 1105–1111

Hickel, J. (2020) *Less Is More: How Degrowth Will Save the World.* London: Random House

Hickel, J. (2019) 'Degrowth: A Theory of Radical Abundance'. *Real-World Economics Review* 87, 54–68.

Hickel, J. and Kallis, G. (2019) 'Is Green Growth Possible?' *New Political Economy* 25(4), 469–486.

Huber, M.T. (2022) *Climate Change as Class War: Building Socialism on a Warming Planet.* London: Verso Books

IEEE (2013) *IEEE Identify the Fourth R-Word in Sustainability: Repair.* [online] available from <https://web.archive.org/web/20160101213644/http://www.ieee.org/about/news/2013/22april_2013.html>

Illich, I. (1973) *Tools for Conviviality*. New York: Harper & Row

IPBES (2019) *Global Assessment Report on Biodiversity and Ecosystem Services* [online] available from <https://ipbes.net/node/35274> [10 February 2023]

Jackson, S.J. (2014) '11 Rethinking Repair'. *Media Technologies: Essays on Communication, Materiality, and Society* 221–239

Jackson, T. (2021) *Post Growth: Life After Capitalism*. Cambridge: Polity

Kallis, G. (2019a) 'Socialism without Growth'. *Capitalism Nature Socialism* 30 (2), 189–206

Kallis, G. (2019b) *Limits: Why Malthus Was Wrong and Why Environmentalists Should Care*. Stanford: Stanford University Press

Kallis, G. (2018) *Degrowth*. Newcastle-Upon-Tyne: Agenda Publishing

Kulwin, N. (2016) 'Elon Musk Wants to Send You to Mars — Here's How He Plans to Do It'. *Vice* [online] 27 September. available from <https://www.vice.com/en/article/59evbk/elon-musk-wants-to-send-you-to-mars-heres-how-he-plans-to-do-it> [6 December 2022]

Laser, S., Paske, A., Estrid, S., Mel, H., Ojala, M., Fehrnenbacher, J., Hepach, M., Celik, L., and Kumar, K.R. (2022) 'The Environmental Footprint of Social Media Hosting: Tinkering with Mastodon'. *EASST Review* [online] 41 (3). available from <https://www.easst.net/article/the-environmental-footprint-of-social-media-hosting-tinkering-with-mastodon/> [26 June 2023]

Latour, B. (2016) 'Fifty Shades of Green'. *Environmental Humanities* 7 (1), 219–225

Lazzarato, M. (1996) 'Immaterial Labour'. in *Radical Thought in Italy: A Potential Politics*. ed. by Virno, P. and H., Michael. Minneapolis: University of Minnesota Press, 133–147

Leigh-Star, S. (1999) 'The Ethnography of Infrastructure'. *American Behavioral Scientist* 43 (3), 377–391

Luxemburg, R. (1951) *The Accumulation of Capital*. London: Routledge & Kegan Paul

Marx, K. (1973) *Grundrisse*. trans. by Nicolaus, M. London: Penguin

Michaux, S.P. (2021) 'The Mining of Minerals and the Limits to Growth'. *Geological Survey of Finland* available from <https://tupa.gtk.fi/raportti/arkisto/16_2021.pdf> [4 April 2023].

Moore, J.W. (2017) 'Metabolic Rift or Metabolic Shift? Dialectics, Nature, and the World-Historical Method'. *Theory and Society* 46 (4), 285–318

Moore, J.W. (2015) *Capitalism in the Web of Life: Ecology and the Accumulation of Capital*. New York and London: Verso Books

Morozov, E. (2014) *To Save Everything, Click Here: The Folly of Technological Solutionism*. London: Penguin

Mueller, S. (2006) *Microprocessors from 1971 to the Present* [online] available from <https://www.informit.com/articles/article.aspx?p=482324&seqNum=2> [25 May 2023]

Oxfam (2020) *Confronting Carbon Inequality: Putting Climate Justice at the Heart of the COVID-19 Recovery* [online] available from <https://oxfamilibrary.openrepository.com/bitstream/handle/10546/621052/mb-confronting-carbon-inequality-210920-en.pdf>

Parks, L. (2009) 'Around the Antenna Tree: The Politics of Infrastructural Visibility'. [6 March 2009] available from <http://www.flowjournal.org/2009/03/around-the-antenna-tree-the-politics-of-infrastructural-visibilitylisa-parks-uc-santa-barbara/> [25 May 2023]

Phillips, L. (2015) *Austerity Ecology & the Collapse-Porn Addicts: A Defence of Growth, Progress, Industry and Stuff*. John Hunt Publishing

Rheingold, H. (1993) *The Virtual Community: Homesteading on the Electronic Frontier*. Reading, MA: Addison-Wesley

Rheingold, H. (1991) *Virtual Reality*. New York: Summit Books

Ritchie, H. and Roser, M. (2021) *Energy Mix* [online] available from <https://ourworldindata.org/energy-mix>

Ritchie, H., Roser, M., and Rosado, P. (2020) 'CO_2 and Greenhouse Gas Emissions'. *Our World in Data* [online] available from <https://ourworldindata.org/co2/country/china> [18 January 2023]

Robbins, P. (2020) 'Is Less More… or Is More Less? Scaling the Political Ecologies of the Future'. *Political Geography* 76, 102018

Saito, K. (2023) *Marx in the Anthropocene: Towards the Idea of Degrowth Communism*. Cambridge: Cambridge University Press

Schiller, D. (1999) *Digital Capitalism: Networking the Global Market System*. Cambridge: MIT Press

Schmelzer, M., Vetter, A., and Vansintjan, A. (2022) *The Future Is Degrowth: A Guide to a World beyond Capitalism*. London and New York: Verso Books

Sharma, R. (2017) *Deep-Sea Mining*. Cham: Springer International Publishing

Soper, K. (2020) *Post-Growth Living: For an Alternative Hedonism*. New York and London: Verso

Srnicek, N. (2016) *Platform Capitalism*. Cambridge: Polity

Starosielski, N. (2015) *The Undersea Network*. Durham and London: Duke University Press

Steffen, W., Richardson, K., Rockström, J., Cornell, S.E., Fetzer, I., Bennett, E.M., Biggs, R., Carpenter, S.R., De Vries, W., and De Wit, C.A. (2015) 'Planetary Boundaries: Guiding Human Development on a Changing Planet'. *Science* 347 (6223). doi: 10.1126/science.1259855

Taffel, S. (2023) 'AirPods and the Earth: Digital Technologies, Planned Obsolescence and the Capitalocene'. *Environment and Planning E: Nature and Space* 6 (1), 433–454

Taffel, S. (2018) 'Hopeful Extinctions? Tesla, Technological Solutionism and the Anthropocene'. *Culture Unbound: Journal of Current Cultural Research* 10 (2), 163–184

Timoney, K.P. and Lee, P. (2009) 'Does the Alberta Tar Sands Industry Pollute? The Scientific Evidence'. *The Open Conservation Biology Journal* [online] 3 (1). available from <https://benthamopen.com/ABSTRACT/TOCONSBJ-3-65> [24 May 2023]

Tyson, M. (2020) *Intel 14nm and AMD/TSMC 7nm Transistors Micro-Compared - CPU - News - HEXUS.Net* [online] available from <https://hexus.net/tech/news/cpu/145645-intel-14nm-amdtsmc-7nm-transistors-micro-compared/> [10 February 2023]

Villegas, A. and Scheyder, E. (2023) 'Chile Plans to Nationalize Its Vast Lithium Industry'. *Reuters* [online] 21 April. available from <https://www.reuters.com/markets/commodities/chiles-boric-announces-plan-nationalize-lithium-industry-2023-04-21/> [26 June 2023]

Vinsel, L. and Russell, A.L. (2020) *The Innovation Delusion: How Our Obsession with the New Has Disrupted the Work That Matters Most*. New York: Currency

Washburn, T.W., Turner, P.J., Durden, J.M., Jones, D.O.B., Weaver, P., and Van Dover, C.L. (2019) 'Ecological Risk Assessment for Deep-Sea Mining'. *Ocean & Coastal Management* 176, 24–39

Wiedenhofer, D., Virág, D., Kalt, G., Plank, B., Streeck, J., Pichler, M., Mayer, A., Krausmann, F., Brockway, P., Schaffartzik, A., Fishman, T., Hausknost, D., Leon-Gruchalski, B., Sousa, T., Creutzig, F., and Haberl, H. (2020) 'A Systematic Review of the Evidence on Decoupling of GDP, Resource Use and GHG Emissions, Part I: Bibliometric and Conceptual Mapping'. *Environmental Research Letters* 15 (6), 063002

Wiedmann, T.O., Schandl, H., Lenzen, M., Moran, D., Suh, S., West, J., and Kanemoto, K. (2015) 'The Material Footprint of Nations'. *Proceedings of the National Academy of Sciences* 112 (20), 6271

Williams, A. and Gilbert, J. (2022) *Hegemony Now: How Big Tech and Wall Street Won the World (And How We Win It Back)*. London: Verso Books

Zalasiewicz, J., Waters, C.N., Summerhayes, C.P., Wolfe, A.P., Barnosky, A.D., Cearreta, A., Crutzen, P., Ellis, E., Fairchild, I.J., and Gałuszka, A. (2017) 'The Working Group on the Anthropocene: Summary of Evidence and Interim Recommendations'. *Anthropocene* 19, 55–60

Zeng, A., Chen, W., Rasmussen, K.D., Zhu, X., Lundhaug, M., Müller, D.B., Tan, J., Keiding, J.K., Liu, L., Dai, T., Wang, A., and Liu, G. (2022) 'Battery Technology and Recycling Alone Will Not Save the Electric Mobility Transition from Future Cobalt Shortages'. *Nature Communications* 13 (1), 1341

Zhao, L., Cheng, Z., Zhu, H., Chen, H., Yao, Y., Baqar, M., Yu, H., Qiao, B., and Sun, H. (2023) 'Electronic-Waste-Associated Pollution of per- and Polyfluoroalkyl Substances: Environmental Occurrence and Human Exposure'. *Journal of Hazardous Materials* 451, 131204

Zuboff, S. (2019) *The Age of Surveillance Capitalism: The Fight for a Human Future at the New Frontier of Power*. New York: Public Affairs

3

BIG CLOUD SOLASTALGIA

Mél Hogan and Gwendolyn Blue

'Digital sustainability' usually pertains to eco-conscious practices that involve modifying existing systems or implementing new strategies to align with sustainability objectives, like reducing carbon emissions, optimizing energy efficiency, and minimizing electronic waste. These can be costly modifications, especially when sustainability goals pertain to large-scale communications and data infrastructures. Too often, initiatives to 'green' or render more sustainable a particular infrastructure are measured against the profit the industry generates in its current state. If the industry is profitable – as cloud computing is – it might lack incentive to change because there are significant financial costs associated with achieving sustainability goals. This in part explains why sustainability goals are often pushed into the future – usually at least a decade or so – to delay costs against profit. This also in part explains why sustainability goals become (economic) discursive distractions, perpetual promises of something better, while little (materially) changes. In practice, 'digital sustainability' in the realm of big tech is largely a type of greenwashing that speaks to data-driven industry initiatives that seek to downplay their role in exacerbating planetary harms. Here, 'planetary harms' refers to natural resource extraction and exploitation, and the use of water, land, and energy to operate the cloud and its ambient fields along various supply chains; but it also refers to the political, affective, and philosophical underpinnings that drive the industry and the market logics now deeply entrenched in uneven, violent, colonial, global, capitalist exchange (Parikka 2011; Gabrys 2011; Maxwell and Miller 2012; Bresnihan and Brodie 2021, 2023; Au 2022; Torres 2024). The best version of sustainability, in this context, would mean radically transforming many interconnected industries on a global scale, abolishing certain projects entirely and reducing corporate profit margins significantly (Pickren 2016, 2017; Brodie 2020; Burrell 2020; Munn 2020; Brodie and Velkova 2021; Lehuedé 2022; Rone 2023).

DOI: 10.4324/9781003441311-4

To illustrate this, companies like Apple, Microsoft, Meta (Facebook), Alphabet (Google), and Amazon each have their corporate sustainability report accessible to the public online that sells us ideas of fair, transparent, clean, modern, energy-efficient futures that will maintain and grow our digital lives, especially (lately) through advancements in 'artificial intelligence' (AI). The foundations and most common uses of 'sustainability' in relation to nature and so-called natural resources in these corporate reports are meant to be about a kind of care towards and acknowledgement of nature's finitudes. If managed well – the idea goes – companies can responsibly extract from nature and replenish it, in perpetuity, to the benefit of all, equally. This is the techno-capitalist promise and pitch: nature can serve us, and our ways of life, forever – but only if left in the hands of the techno-capitalists making those promises. In short, big tech (increasingly as big cloud) companies position themselves as the best custodians of nature and of natural resources.[1] How did this come to be?

Meta's 2023 Sustainability Report quotes Mark Zuckerberg who says, "The possibilities our technology will unlock for people only matter if we have a safe and thriving planet."[2] Here, Zuckerberg situates Meta as being best positioned to make 'a safe and thriving planet' happen:

> We see our role as protecting people and the planet through responsible operations — minimizing our emissions and the energy and water used to power our data centers that enable users to access our products and the workplaces where those products are built and managed — while protecting workers and the environment in our supply chain.
>
> *(2023: 7)*

The promise here is that Meta will self-regulate because it has 'environmental politics' embedded in its business model, as a core value. Microsoft's 2022 Sustainability Report notes that it announced in 2020 that it "will be a carbon negative, water positive, zero waste company that protects ecosystems—all by 2030."[3] In their report, Microsoft also promises to innovate with thermal energy, measure their emissions, decarbonize its operations, to be 'water positive,' to protect land, to recycle its hardware, etc., all with the goal of growing data markets. Like Microsoft, the Google 2023 Environmental Report promises to make its company, other industries, and individuals, all more resilient via various sustainable technological innovations.[4] Google seems particularly invested in the idea of using predictive 'AI' to help individuals make better choices about the products they consume and help industry become less wasteful and more efficient in the products they sell. This maintains current business models, though this renewed focus on efficiency is likely to lead to *more* production, consumption, and growth – as designed. Amazon's 2022 Sustainability Report squarely defines itself as a technofix: "Amazon was built on the belief that with understanding, ingenuity, and innovation, we can more effectively overcome any challenge we face. We believe addressing environmental and societal challenges requires the same mindset" (2022: 2). Like Amazon, Apple holds as its company ethos that the idea of business itself – of capitalism – is *not* a problem: "we demonstrated that the choice between a thriving business and a thriving planet is a false one" (2022: 3).[5] Like Microsoft, Apple's 2022 Environmental Progress Report also aims for a better world by 2030 – a round number that especially appeals to an engineering mindset of precision, predictability and calculability.

These are all excerpts that reflect the corporate social responsibility discourse we've become accustomed to (from the United States especially, where these companies are based) – discourse that companies must have as part of their brand, and that admittedly (can) sound compelling if the goal is a 'greener' future. But, these excerpts also work to demonstrate how much the cloud industry covers up the larger social and political terrains upon which their projects are founded, as their loyalties are strictly with shareholders. Sustainability is about maintaining an image and an ideal about the future – but rarely does it account for the social, emotional, and affective labour involved in the transition. Specifically, when one thinks about sustainability, how far back into the past does one travel? Surely we're not trying to sustain the planetary conditions that we have today. Instead we must imagine a 'pre-time,' when things were in a kind of harmony or balance, a starting point from which things could be exchanged rather than abused and exploited. But was there ever such a time? Or is the very idea of sustainability itself, in fact, enabling ongoing planetary abuses and exploitation?

These questions tend to mean that the work of unearthing the social and political terrains of big tech is left to critical scholars (and activists, journalists, artists) who can understand digital sustainability as a complicated interplay between promises, visions, capitalism and power – not just a technological fix that awaits funds to then be implemented by engineers.

Discussing the environmental impacts of data centres and the cloud is important, but so too, increasingly, is acknowledging the felt conditions of this orientation to research. What purpose does it serve to make known the energy consumption of data centres, how much water 'AI' consumes, or the extractivist nature of the cloud, if the topic does not resonate at an existential level? Because the materialities of big tech (data centres, cable landing sites, energy sources, etc.) are often out of reach for researchers to access directly, 'the cloud' becomes an important affective and conceptual site and object of inquiry that invites myriad approaches to understand, analyse, and ideally (one day) remake or abolish it (Velkova 2016; Veel 2017; Taylor 2021; Narayan 2022). Particularly, in the context of sustainability initiatives, what big tech can't admit to, let alone anticipate or solve, is the way environmental collapse registers in our bodies and psyches – as the protesters, programmers, influencers, engineers, scholars, users, citizens, scientists, luddites, and other people who pay with our data to the benefit of an industry.

So, while sustainability has become popular shorthand, generally, for the attention paid to environmental impacts and to the idea that natural resources are by definition finite, the concept of sustainability is a contested idea. The concept carries often implicit and unacknowledged claims that are entangled with all kinds of problematic power relations, rooted in longstanding settler-colonial logics. However, when power relations are surfaced, they are often presented in myriad ways as a tension between *extraction* and *overuse*, be that with any or all natural resources, like water, salt, silver, or wood, etc. The idea is that scarcity is an undesirable outcome that requires oversight, planning, mitigation, and management – that extraction is possible but requires replenishing as an integral part of the processes of sustaining and/or for sustenance. If the goal is simply replenishing what was taken, we become stuck in a mode of perpetual extractive colonialism. We see resonances of this logic with 'data extraction' and 'data colonialism' (Couldry and Mejias 2019), especially in light of recent cases calling out companies, like OpenAI, on data-scraping the internet in order to feed proprietary large language models (as 'AI').[6] Arguably, if a company (or industry writ large) generally accepts the extraction of so-called primary natural resources as a model, it's likely that similar practices will be extended to such assets as data, where the internet is the commons to be plundered for profit. Data extractivism is at the service of capitalism, neoliberalism, fascism, settler-colonialism and corporate surveillance.

What is made less obvious through corporate sustainability-as-goal are the permanent disruptions and their consequences, the hard limits to replacement after exploitation, the dependencies created, and the ways in which imagination itself is limited by these often permanent alterations. These are all deeply entrenched profit-driven decisions, but they are by no means the only way for humanity to survive on this planet. As David Graeber famously noted in *The Utopia of Rules: On Technology, Stupidity, and the Secret Joys of Bureaucracy* "the ultimate, hidden truth of the world is that it is something that we make, and could just as easily make differently" (2015: 89). While we've gone so far into normalizing capitalist exploits, even at the expense of our own living conditions, there are other futures possible. What if we were to take this world-(re)making plea seriously and imagine the fate of our planet? And for the sake of this chapter, how might we imagine the planet as something untethered from big cloud? Or, ways that 'easily make differently' a world that isn't about inevitable collapse exacerbated by data-driven ideals? How do we unmake this (unsustainable) version of a data world?

Solastalgic

Before moving forward with this, it's important to note that we are not equal, neutral observers, or thinkers, or feelers, uninfluenced or unmoved by the environment which we inhabit (Bladow and Ladino 2018; Fuller and Goriunova 2019; Estok 2019). As environmental conditions change – lately for the worse, often without warning – so have we. And with that, so has our ability, will, and desire to reflect on those very conditions. We might not know exactly how the trauma, indifference, stress, sadness, or denial of a changing planet is affecting our ability to want or to make new worlds. But it's clear that feeling more beaten down by pandemics and precarity, inhaling smoke from wildfires, enduring loss and grief, and the facing doom and gloom of disasters like floods and hurricanes renders the conditions for making a new world more difficult; and while these problems, and the fears they invoke in us, grow ever more urgent, urgency itself can't promise results. In fact, there's more evidence that humans surrender quickly to new conditions (even horrible ones), than push for change. This is known as "shifting baselines syndrome" where humans don't remember what they've lost, they simply adjust to what is, as a form of amnesia.[7] This syndrome is all the more worrisome when things promise to get a lot worse, more quickly, in the next decade (Wallace-Wells 2022).[8] A study published in 2019 in the Proceedings of the National Academy of Sciences showed that people learn to accept extreme climate events as 'normal' in as little as two years (Moore et al. 2019). In that study, researchers show that human conceptions of normal or abnormal are influenced by a range of factors including "generational turnover, memory limitations, and cognitive biases" (Moore et al. 2019: 4905). This means that we cannot expect to feel or endure the future of climate in ways that are necessarily collective, even when changes are happening in a specific location, in real-time. Humans will always have an ability and impulse to narrate their experience in ways that assuage feelings that are unpleasant to them specifically. As we've seen with the COVID-19 pandemic, there is no guarantee of a shared experience of trauma that leads to effective collective action. The point here is not that we should all feel and respond in the same way to catastrophe, but that sharing a sense of a particular problem enables more possibilities for remaking worlds with different endings.

Whether felt individually or collectively, this particular anxiety has a name: 'solastalgia.' Solastalgia is defined by psychologists as "the experience of distress from belonging to a home that is undergoing change"[9] – this planet is that 'home.' In *Earth Emotions: New Words for a New World* (2019), Glenn A. Albrecht's explains that 'solastalgia' was coined by him in the 2000s specifically to name what we are experiencing, "by the combination of the Latin words *sōlācium* (comfort) and the Greek root *-algia* (pain, suffering, grief), that describes a form of emotional or existential distress caused by environmental change."[10] On *PubMed*, a more medicalized interpretation of 'solastalgia' is contrasted to 'nostalgia,' where 'nostalgia' is "the melancholia or homesickness experienced by individuals when separated from a loved home," and "solastalgia is the distress that is produced by environmental change impacting on people while they are directly connected to their home environment."[11] We may be nostalgic for what was or might have been, but we are solastalgic in our embodiments of the present crisis. The point here is that we can no longer write from a pre-pandemic or pre-climate catastrophe time, or as though the stressors we now face are something we will ever return from – we have to start situating our writing within the affective register of solastalgia. We might not be aware that we have been transformed by global crises, but we have.

So, what if we consider solastalgia the condition of (and context for) thinking about big cloud? What if, in fact, big cloud logics are already constituted of solastalgia? Might this help explain the overly engineered approach to future-building that has failed and keeps failing while celebrating itself and offering itself as the saviour and solution to all our problems?

Vibe check

When we look at proposals made by scholars, activists, artists, and practitioners, either directly or through critiques of the status quo (via big tech, as big cloud) we can assess what's being communicated below the surface, and how worlds are (re) made.

The best word for this is 'vibes' or 'vibe check,' as it's understood to capture the general feel of something; a read. We turn briefly to 'vibes' here, as a conceptual framework, to make a case that new imaginaries are often contained within certain *affects* and *feelings* like solastalgia, but that necessarily radiate outwardly. *Vibes* refers to an "emotional state or the atmosphere of a place as communicated to and felt by others"[12] and "a distinctive feeling or quality capable of being sensed."[13] A great deal of work has been done by critical media scholars and others to delineate the vast problems of 'the cloud' (and by extension, 'AI'); others have already assessed the environmental impacts of data centres, tracked e-waste, located trespassing undersea cables, identified problematic energy sources, mining practices and labour conditions, and so on. Rather than reiterating the important work already outlining these impacts (Gabrys 2011; Parikka 2011; Maxwell and Miller 2012; Blum 2013; Mosco 2014; Hu 2015; Parks and Starosielski 2015; Peters 2015; Starosielski 2015, 2021; Stengers 2016; Cubitt 2017; Amoore 2020), it seems more fruitful to assess the feeling/affect that solastalgia encapsulates, and what that affect's presence in our collective imagination tells us about the new worlds we imagine (or imagine lost).

By now, the social and environmental harms of the cloud and its undergirding infrastructures are well-documented.[14] Among many others, Steven Gonzalez explains that the cloud is fuelled by "dirty" electricity grids (2022), James Glanz reports on how a data centre became a shelter during a hurricane (2017), Ingrid Burrington maps infrastructure (2015), Dwayne Monroe shows how the cloud became the computational site of global capitalism (2022), Dan Greene shows us who our internet landlords are (2022), Jeffrey Moro demonstrates the impacts of air-conditioning the internet (2021), David Gray Widder, Sarah West and Meredith Whittaker publicly denounce the concentration of computing power (2023), Anne Pasek shows us that carbon offsets are a scam (2019), Shaolei Ren and collaborators demonstrate AI's enormous water usage (2023), Asta Vonderau writes about the effects of migration of the global cloud infrastructure towards the Northern hemisphere (2019) – together, these help form a giant network of compelling arguments against big cloud's continued abuses of the planet and its inhabitants. What scholars (as well as artists, journalists, activists) demonstrate, as a collective critique, is that the social imaginary of big cloud has convinced us that salvation comes through data; having enough data to predict the future, and using those futures to imagine new worlds. But who is present in these imagined future worlds and under what new (data) conditions?

The scale of big cloud's impact is massive. Apple, Microsoft, Meta, Alphabet, and Amazon are the largest internet companies worldwide, with billions of users and a combined market value of almost 7 trillion U.S. dollars.[15] Specifically, Amazon Web Services (AWS) (34%), Microsoft Azure (22%) and Google Cloud Platform (9.5%) are the largest providers of cloud computing in the world.[16]

Meta envisions a future where cloud computing plays a central role in a 'metaverse,' a concept raising concerns about data privacy, surveillance, and monopolistic control over virtual spaces. Amazon's relentless expansion of AWS can be seen as concentrating too much on power in the hands of one company, potentially limiting competition and innovation in the cloud industry. Google Cloud often faces criticism for its data collection practices and potential misuse of user information. Apple's emphasis on privacy is recognized as a marketing tactic, and their walled garden approach limits interoperability. Microsoft Azure's hybrid cloud strategy locks businesses into their ecosystem. Data power is concentrated with these five big cloud companies, in their infrastructures, which play a disproportionate role in shaping future worlds – this is much less about sustaining and much more about forging new modalities to coexist with, if not *as*, data. In this sense, big cloud is not only a communications infrastructure but the holder of potentialities – we have relinquished all of our nodes as data points to the prediction industry, and have taken a backseat. Now, we wait and see what it will deliver. Will it keep the promises these companies made in their sustainability reports?

If it were possible to do a 'vibe check' on a somewhat comprehensive review of the critical literature on data centres and the cloud, it would reveal big cloud as the end product of all our efforts, the inevitability of a particular kind of settler-colonial logic strengthened by the rise of alt-right politics, white supremacy, and fascism, which have historically been a catalyst and response to changing living conditions, when people get scared and attempt to save themselves (Dauvergne 2020; Crawford 2021; O'Gieblyn 2021; Bridle 2022; Lovink 2022; McQuillan 2022). Whatever the case may be, never before have we been beholden to such a technological apparatus; a 'Cloud Empire' run by despots (Lehdonvirta 2024).

Remade worlds

In 2014, Karin Ljubič Fister and Iztok Fister Jr. contacted two biotechnologists from the Biotechnical Faculty at the University of Ljubljana, Prof Dr Borut Bohanec and Dr Jana Murovec, with an idea about plant-based data storage.[17] As a collaborative effort, the scientists prepared a DNA-encoded computer programme and injected it into a *Nicotiana benthamiana* plant (a close relative of tobacco).[18] On the project website, they explain:

> We wrote a program in [*the*] Python programming language, transformed it from ASCII to binary and translated the sequence of 0s and 1s into A, C, T, G sequence of nucleotides. Our code DNA was incorporated into *N. benthamiana* plant and today we have our computer program stored in its growing progeny and seeds. The incorporated data was obtained from a leaf with 100 percent accuracy. It is the first practical utilization of storing meaningful data in multicellular organisms.[19]

They cloned plants injected with their DNA programme. By planting a cutting of the altered leaf, and planting it, they were able to produce a new plant in which all the leaves and seeds contained the programme encoded in the original plant's DNA. In an interview with BLDGBLOG, Fister explains her vision and philosophy. She says, "all of the archives in the world could be stored in one box of seeds," suggesting that all the world's data could be stored in a seed vault for thousands of years, as "seed drives, not hard drives."[20] The stated idea is that current storage modalities are not sustainable. But the critique is far deeper than this; it is about creating a distance between *memory* and *archive* as we hand over our data to corporate third parties, and about a distance between *individual memory* and *shared context*, as we determine storage to be storable at all in bits and bytes (Birhane 2018).[21] Big cloud might be better understood as a kind of pre- but inevitably- 'dead media,' – though not so much in the sense Bruce Sterling (2020) posits, which is "a compilation of obsolete and forgotten communication technologies,"[22] but more so as something lacking in 'vital materialism,' as Jane Bennett (2010) might put it.[23] Or, as something in between, as Garnet D. Hertz and Jussi Parikka (2011) explain, the cloud (as a network of media) ultimately will fail and leak toxically for many human lifespans, as 'zombie media.'[24] Each of these theoretical concepts is shaped in part by solastalgia. Because if we play out the ends of sustainability and take those consequences seriously, there is no cloud or 'AI.' There will be mere remnants of digital data infrastructure in the sacrifice zone that is the entire planet, long past the limits of sustainability. It will outlive its purpose and remain as a kind of "immortal media" – non-functioning, but material.[25] Even post-sustainability – as raw materials are completely depleted – there will be evidence of this attempt to encapsulate ourselves in data.

Cyrus Clarke of the Grow Your Own Cloud (GYOC) artist collective addresses the idea of "Data Immortality" in a 2021 *Medium* post, saying "current methods of archiving data face the problems of obsolescence, reliance on scarce materials, changing formats, decay and accumulating piles of e-waste."[26] And, like Fister years earlier, Clarke turns to DNA. Clarke writes:

> DNA on the other hand lasts a remarkably long time, is incredibly stable, and is found in abundance everywhere. Nature has also demonstrated that DNA can

> last for thousands of years. A study of DNA extracted from the leg bones of extinct moa birds in New Zealand found that its half-life is 521 years.[27]

And like Fister, GYOC turned to plants – and built a plant shop – as an alternative to data centres. Other artists have, too. Kyriaki Goni proposes a garden because, as her project's curatorial text states, "In the age of climate crisis, the constant production, accumulation, and storage of data raise important questions regarding their impact on the natural environment."[28] The project reflects on possible answers:

> As the storage space changes from the "cloud" to the earth, and as control passes from the companies to the users, the life [*circle*] of data follows that of a plant, fostering a relation of interdependence and care. In a peculiar garden, users become the plants' gardeners, whereas plants respectively become the gardeners of the stored information.[29]

In each of these projects, there is a longing to preserve and be preserved. That longing is in itself, arguably, solastalgic. We argue that these projects are solastalgic because they do not in fact propose viable new solutions to current problems. Instead, they propose alternatives that in themselves draw more attention to solastalgia. For example, solastalgia is manifested in this headline from *ZDNet*: "With the rise of global instability, can ceramic storage save our digital culture?"[30] Here the implication is that we need materials and concepts that will outlast us, humans, and a turn to ceramic, like DNA, holds promise for this. The idea that ceramic is "impervious to water, chemicals, and radiation; it's emboldened by fire"[31] captures the sentiment. Microsoft, investing in this project explains that it is a "five-dimensional (5D) data storage" that "uses molecule-sized nanostructures created in silica glass to store information," and is "10,000 times denser in storage than a Blu-Ray disc."[32] It can store data for "up to 13.8 billion years."[33] Here, the imaginary is that the planet lives on for a very long time – too long – and that technology and humans might come and go, but that a record of humanity remains in circulation. To make records that outlast scientific estimates that in "about a billion years the sun will become hot enough to boil our oceans"[34] which spells the end of humanity, with or without climate change, is fascinating. But perhaps understanding that humans and all life on earth will eventually and inevitably die off, that the planet itself is not eternal no matter how well we tend to it, helps explain both the feeling that we need to make the most of it while it's here, and the defeat of knowing it won't last no matter what we do. That being said, on the human scale, a billion years is a long time to wait out the end.

In writing this chapter, we discovered many other concepts that attempt to situate the range of feelings we have about our relationship to the planet: *soliphilia* as "the love of the totality of our place relationships, and a willingness to accept the political responsibility for protecting and conserving them at all scales" (Albrecht

2019: 121) ; *tierratrauma* as "a sudden negative environmental experience bearing long-term consequences" (Albrecht 2019: 84); *terrafurie* as "the rage of those who witness the destructive compulsions of techno-industrial civilization" (Albrecht 2019: 85); *sumbiotude* as "contemplation and completion of a lifespan with the loving companionship of humans and non-humans";[35] or *noctalgia*, meaning "sky grief" from the loss of darkness due to light pollution.[36] Courtney O'Dell Chaib reminds us that "cultivating affinity and attachment within ecological destruction requires thinking through how so-called 'negative' affects like disgust, revulsion, melancholy, shame, and despair are important parts of ecological theory and activism."[37]

Materials like DNA, glass, and ceramics show promise for data storage but also for dealing with the grief and rage embedded in many of the concepts emerging to describe our existence on earth. DNA offers high data density and longevity, and this quells our fears of not being remembered. Glass storage uses etched glass plates for durability, which appeases our sense that floods and fires might erase us. Ceramic storage, especially 3D-printed structures, offers robustness and long-term data retention, which makes similar promises but also offers a return to cuneiform ideals, and lends simplicity to our survival by way of what we leave behind. All of these expose how vulnerable the data centre is – the frailty of networked hard drives in catastrophic times (Bonde Thylstrup 2023).

The relentless growth of data centres has come to symbolize our insatiable thirst for data and for survival, and this presents a dire environmental challenge if 'making new worlds' isn't taken seriously. The cloud – these data centres – are major contributors to escalating energy consumption and carbon emissions and pose a looming threat to our planet. They make us feel worse about the future generally, because greenwashing functions more like grooming and gaslighting this far into global boiling. Without urgent and substantial change, a future where data centres continue to consume vast amounts of energy, often derived from fossil fuels, will worsen climate change, increase greenhouse gas emissions, and further strain ecosystems. Resource-intensive data centre expansion, including water usage, may trigger social and political conflicts and trigger ecological imbalances at scales we've not yet witnessed. Neglecting the imperative to remake the world may mean foregoing the digital age's potential, because so far, environmental degradation is the longest-lasting legacy of technological progress.

Notes

1 https://www.irishexaminer.com/opinion/commentanalysis/arid-41226262.html.
2 https://sustainability.fb.com/2023-sustainability-report/.
3 https://www.microsoft.com/en-us/corporate-responsibility/sustainability/report.
4 https://sustainability.google/reports/google-2023-environmental-report/.
5 https://www.apple.com/environment/pdf/Apple_Environmental_Progress_Report_2023.pdf.

6 https://www.washingtonpost.com/technology/2023/06/28/openai-chatgpt-lawsuit-class-action/.
7 https://www.vox.com/energy-and-environment/2020/7/7/21311027/covid-19-climate-change-global-warming-shifting-baselines.
8 https://www.nytimes.com/interactive/2022/10/26/magazine/climate-change-warming-world.html.
9 "A Psychologist Offers 3 Tips to Deal with 'Solastalgia'" https://www.forbes.com/sites/traversmark/2023/09/14/a-psychologist-offers-3-tips-to-deal-with-solastalgia/.
10 https://en.wikipedia.org/wiki/Solastalgia.
11 https://pubmed.ncbi.nlm.nih.gov/18027145.
12 https://www.encyclopedia.com/humanities/dictionaries-thesauruses-pictures-and-press-releases/vibe-0.
13 https://www.merriam-webster.com/dictionary/vibe.
14 https://commonplace.knowledgefutures.org/pub/jpy7pbq0/release/1.
15 https://www.statista.com/topics/4213/google-apple-facebook-amazon-and-microsoft-gafam/#topicOverview.
16 https://www.knowledgehut.com/blog/cloud-computing/top-cloud-computing-companies.
17 Falling Walls Lab 2015 – Karin Ljubic Fister – Breaking the Wall of Data Storage. https://www.youtube.com/watch?v=d2eDY30hQ8s&ab_channel=FallingWallsFoundation.
18 http://www.storing-data-into-living-plant.net/experiment.
19 http://www.storing-data-into-living-plant.net/experiment.
20 Landscapes of Data Infection https://bldgblog.com/2016/02/landscapes-of-data-infection/.
21 https://aeon.co/ideas/descartes-was-wrong-a-person-is-a-person-through-other-persons.
22 https://www.wired.com/beyond-the-beyond/2020/05/dead-media-beat-dead-media-twenty-one-years-ago/.
23 https://read.dukeupress.edu/books/book/1346/Vibrant-MatterA-Political-Ecology-of-Things
24 Five Principles of Zombie Media https://www.researchgate.net/publication/273062231_Five_Principles_of_Zombie_Media
25 https://blog.dshr.org/2014/06/more-on-long-lived-media.html
26 https://medium.com/@growyourowncloud/all-the-worlds-data-in-a-single-drop-df0275dc09e0
27 https://medium.com/@growyourowncloud/all-the-worlds-data-in-a-single-drop-df0275dc09e0.
28 https://www.onassis.org/whats-on/data-garden.
29 https://www.onassis.org/whats-on/data-garden.
30 https://www.zdnet.com/article/ceramic-storage-will-save-our-digital-culture.
31 https://www.theatlantic.com/technology/archive/2017/01/human-knowledge-salt-mine/512552.
32 https://www.microsoft.com/en-us/research/project/project-silica/.
33 https://www.theverge.com/2016/2/16/11018018/5d-data-storage-glass.
34 https://theconversation.com/the-sun-wont-die-for-5-billion-years-so-why-do-humans-have-only-1-billion-years-left-on-earth-37379.
35 https://theconversation.com/sumbiotude-a-new-word-in-the-tiny-but-growing-vocabulary-for-our-emotional-connection-to-the-environment-136616.
36 https://www.space.com/light-pollution-loss-dark-skies-noctalgia.
37 https://journal.equinoxpub.com/JSRNC/article/view/18439/25573.

References

Albrecht, G.A. (2019). *Earth Emotions: New Words for a New World.* Ithaca, NY: Cornell University Press. ISBN 9 7815 0171 5228.

Amoore, L. (2020). *Cloud Ethics: Algorithms and the Attributes of Ourselves and Others*. Durham, NC: Duke University Press.

Au, Y. (2022). Data Centres on the Moon and Other Tales: A Volumetric and Elemental Analysis of the Coloniality of Digital Infrastructures. *Territory, Politics, Governance*. https://doi.org/10.1080/21622671.2022.2153160.

Bladow, K. & Ladino, J. (2018). *Affective Ecocriticism: Emotion, Embodiment, Environment*. Lincoln, NE: University of Nebraska Press.

Blum, A. (2013). *Tubes: A Journey to the Center of the Internet* (Reprint edition). New York, NY: Ecco.

Bonde Thylstrup, N. (2023). The World's Digital Memory Is at Risk. *New York Times* (21 June) https://www.nytimes.com/2023/06/21/opinion/digital-archives-memory.html (Accessed 13 November 2023).

Bresnihan, P. & Brodie, P. (2021). New Extractive Frontiers in Ireland and the Moebius Strip of Wind/Data. *Environment and Planning E: Nature and Space* 4(4), 1645–1664. https://doi.org/10.1177/2514848620970121.

Bresnihan, P. & Brodie, P. (2023). Data Sinks, Carbon Services: Waste, Storage and Energy Cultures on Ireland's Peat Bogs. *New Media & Society* 25(2), 361–383.

Bridle, J. (2022). *Ways of Being: Animals, Plants, Machines: The Search for a Planetary Intelligence*. New York, NY: Farrar, Straus and Giroux.

Brodie, P. (2020). Climate Extraction and Supply Chains of Data. *Media, Culture and Society*, 42(7–8), 1095–1114. https://doi.org/10.1177%2F0163443720904601.

Brodie, P. & Velkova, J. (2021). Cloud Ruins: Ericsson's Vaudreuil-Dorion Data Center and Infrastructural Abandonment. *Information, Communication and Society* 24(6) #AoIR2020 Special Issue, 869–885. https://doi.org/10.1080/1369118X.2021.1909099.

Burrell, J. (2020). On Half-Built Assemblages: Waiting for a Data Center in Prineville, Oregon. *Engaging Science, Technology, and Society* 6, 283–305.

Burrington, I. (2015). How Railroad History Shaped Internet History. *The Atlantic* (24 November) https://www.theatlantic.com/technology/archive/2015/11/how-railroad-history-shaped-internet-history/417414/ (Accessed 13 November 2023).

Couldry, N. & Mejias, U.A. (2019). Data Colonialism: Rethinking Big Data's Relation to the Contemporary Subject. *Television & New Media* 20(4), 336–349. https://doi.org/10.1177/1527476418796632.

Crawford, K. (2021). *Atlas of AI: Power, Politics, and the Planetary Costs of Artificial Intelligence*. New Haven, CT: Yale University Press.

Cubitt, S. (2017). *Finite Media: Environmental Implications of Digital Technologies*. Durham, NC: Duke University Press.

Dauvergne, P. (2020). *AI in the Wild: Sustainability in the Age of Artificial Intelligence*. Cambridge, MA: The MIT Press.

Estok, S.C. (2019). Ecophobia, the Agony of Water, and Misogyny. *ISLE: Interdisciplinary Studies in Literature and Environment* 26(2), 473–485. https://doi.org/10.1093/isle/isz049.

Fuller, M. & Goriunova, O. (2019). *Bleak Joys: Aesthetics of Ecology and Impossibility*. Minneapolis, MN: University of Minnesota Press.

Gabrys, J. (2011). *Digital Rubbish: A Natural History of Electronics*. Ann Arbor: University of Michigan Press.

Glanz, J. (2017). How the Internet Kept Humming During 2 Hurricanes. *The New York Times* (18 September) https://www.nytimes.com/2017/09/18/us/harvey-irma-internet.html (Accessed 13 November 2013).

Gonzalez Monserrate, S. (2022). The Staggering Ecological Impacts of Computation and the Cloud. *The MIT Press Reader* (14 February) https://thereader.mitpress.mit.edu/the-staggering-ecological-impacts-of-computation-and-the-cloud/ (Accessed 13 November 2023).

Graeber, D. (2015). *The Utopia of Rules: On Technology, Stupidity, and the Secret Joys of Bureaucracy*. New York, NY: Melville House.

Greene, D. (2022). Landlords of the Internet: Big Data and Big Real Estate. *Social Studies of Science* 52(6), 90427. https://doi.org/10.1177/03063127221124943.

Hu, T.H. (2015). *A Prehistory of the Cloud*. Cambridge, MA: The MIT Press.

Lehdonvirta, V. (2024). *Cloud Empires: How Digital Platforms Are Overtaking the State and How We Can Regain Control*. Cambridge, MA: The MIT Press.

Lehuedé, S. (2022). Territories of Data: Ontological Divergences in the Growth of Data Infrastructure. *Tapuya: Latin American Science, Technology and Society* 5(1). 10.1080/25729861.2022.2035936.

Li, P., Yang, J., Wierman, A. & Ren, S. (2023). Towards Environmentally Equitable AI via Geographical Load Balancing. *UC Riverside* (27 June). https://escholarship.org/uc/item/79c880vf (Accessed 13 November 2023).

Lovink, G. (2022). *Extinction Internet*. https://networkcultures.org/blog/publication/extinction-internet/ (Accessed 13 November 2023).

Maxwell, R. & Miller, T. (2012). *Greening the Media* (1st edition). New York, NY: Oxford University Press.

McQuillan, D. (2022). *Resisting AI: An Anti-fascist Approach to Artificial Intelligence* (1st edition). Bristol, UK: Bristol University Press.

Moore, F.C., Obradovich, N., Lehner, F. & Baylis, P. (2019). Rapidly Declining Remarkability of Temperature Anomalies May Obscure Public Perception of Climate Change. *Proceedings of the National Academy of Sciences* 116(11), 4905–4910. https://www.pnas.org/doi/epdf/10.1073/pnas.1816541116.

Monroe, D. (2022). Seeding the Cloud. *Logic(s) Magazine* (27 March). https://logicmag.io/clouds/seeding-the-cloud/ (Accessed 13 November 2023).

Moro, J. (2021). Air-Conditioning the Internet: Data Center Securitization as Atmospheric Media. *Media Fields Journal: Critical Explorations in Media and Space* (26 April). http://mediafieldsjournal.org/air-conditioning-the-internet/2021/4/26/air-conditioning-the-internet-data-center-securitization-as.html (Accessed 13 November 2023).

Mosco, V. (2014). *To the Cloud: Big Data in a Turbulent World*. New York, NY: Routledge.

Munn, L. (2020). Injecting Failure: Data Center Infrastructures and the Imaginaries of Resilience. *The Information Society*, 36(3), 167–176. https://doi.org/10.1080/01972243.2020.1737607.

Narayan, D. (2022). Platform Capitalism and Cloud Infrastructure: Theorizing a Hyper-scalable Computing Regime. *Environment and Planning A: Economy and Space*. https://journals.sagepub.com/doi/full/10.1177/0308518X221094028.

O'Gieblyn, M. (2021). *God, Human, Animal, Machine: Technology, Metaphor, and the Search for Meaning*. New York, NY: Doubleday.

Parikka, J. (2011). *Medianatures: The Materiality of Information Technology and Electronic Waste*. London, UK: Open Humanities Press.

Parks, L. & Starosielski, N. (Eds.). (2015). *Signal Traffic: Critical Studies of Media Infrastructures*. Champaign: University of Illinois Press.

Pasek, A. (2019). Managing Carbon and Data Flows: Fungible Forms of Mediation in the Cloud. *Culture Machine* (2 April). https://culturemachine.net/vol-18-the-nature-of-data-centers/managing-carbon/.

Peters, J.D. (2015). *The Marvelous Clouds: Toward a Philosophy of Elemental Media*. Chicago, IL; London, UK: University of Chicago Press.

Pickren, G. (2016). 'The Global Assemblage of Digital Flow': Critical Data Studies and the Infrastructures of Computing. *Progress in Human Geography* 42(2), 225–243.

Pickren, G. (2017). The Factories of the Past Are Turning Into the Data Centres of the Future. *Imaginations* 8(2), Location and Dislocation: Global Geographies of Digital Data, 22–29. https://journals.library.ualberta.ca/imaginations/index.php/imaginations/article/view/29366.

Rone, J. (2023). The Shape of the Cloud: Contesting Data Centre Construction in North Holland. *New Media & Society*. https://doi.org/10.1177/14614448221145928.

Starosielski, N. (2021). *Media Hot and Cold*. Durham, NC: Duke University Press Books.

Starosielski, N. (2015). *The Undersea Network*. Durham, NC: Duke University Press Books.

Stengers, I. (2016). *In Catastrophic Times: Resisting the Coming Barbarism*. London, UK: Open Humanities Press.

Taylor, A.R.E. *(2021). Future-Proof: Bunkered Data Centres and the Selling of Ultra-Secure Cloud Storage. *Journal of the Royal Anthropological Institute* 27, 76–94. https://rai.onlinelibrary.wiley.com/doi/10.1111/1467-9655.13481.

Torres, É.P. (2024). *Human Extinction: A History of the Science and Ethics of Annihilation*. New York, NY: Routledge. ISBN 9781032159065.

Travers, M. (2023). A Psychologist Offers 3 Tips To Deal With 'Solastalgia.' *Forbes* (14 September). https://www.forbes.com/sites/traversmark/2023/09/14/a-psychologist-offers-3-tips-to-deal-with-solastalgia/ (Accessed 13 November, 2023).

Veel, K. (2017). Uncertain Architectures—Performing Shelter and Exposure. *Imaginations* 8(2): Location and Dislocation: Global Geographies of Digital Data, 30–41. https://journals.library.ualberta.ca/imaginations/index.php/imaginations/article/view/29367.

Velkova, J. (2016). Data That Warms: Waste Heat, Infrastructural Convergence and the Computation Traffic Commodity. *Big Data & Society* 3(2). https://doi.org/10.1177/2053951716684144.

Vonderau, A. (2019). Storing Data, Infrastructuring the Air: Thermocultures of the Cloud. *Culture Machine* (2 April). https://culturemachine.net/vol-18-the-nature-of-data-centers/storing-data/.

Wallace-Wells, D. (2022). Beyond Catastrophe: A New Climate Reality Is Coming Into View. *The New York Times* (26 October). https://www.nytimes.com/interactive/2022/10/26/magazine/climate-change-warming-world.html (Accessed 13 November 2023).

Widder, D.G., West, S. & Whittaker, M. (2023). Open (For Business): Big Tech, Concentrated Power, and the Political Economy of Open AI. SSRN Scholarly Paper. Rochester, NY (17 August). https://doi.org/10.2139/ssrn.4543807.

4

OPERATIVE LANDSCAPES OF DIGITISATION, COLLATERAL LANDSCAPES OF CIRCULARITY

Stephen Cornford

In the Chilean mining town of Chuquicamata (Figure 4.1), labourers and their families are being relocated. The settlement is dwarfed by the scale of the eponymous neighbouring copper mine, smaller even than the tailing ponds to which waste water is pumped. After retirement, pensioners from the mine typically live for only three to four years due to silicosis caused by the volume of dust inhaled during their employment. In the opening pages of *Scorched Earth*, Jonathan Crary writes of the inability of digital technics to moderate its consumption and metabolise its detritus, describing: "a world operating without pause, without the possibility of renewal or recovery, choking on its own waste" (2022: 2). In Chuquicamata, in the shadow of a mine operating 24/7, retired workers literally choke on the waste of connectivity. In what follows, I broach the unsustainability of the digital by extending the opposition Crary establishes between, on the one hand, ceaseless operativity and, on the other, an inability to remediate the resulting waste: an opposition between operative and collateral landscapes.

Digital technologies are made from the landscape, their construction re-shapes that landscape, and it is the effects of mining on landscape that primarily concern me here. From one perspective, the question of digital sustainability can be boiled down to this relationship between technology and landscape. How has the digitisation of our tools and technics impacted the planetary and ecological systems that enable abundant biodiversity, hydration, oxygenation, and habitation? Or, to frame the question in the implicit circularity of the spherical: can the biosphere, lithosphere, and atmosphere keep pace with the remediation required of them by the technosphere? The question is hardly new. In 1971, Lewis Mumford described industrialisation as "unlike organic systems" in that it "has no built-in method of controlling its growth or modulating the enormous energy it commands in order to maintain a dynamic equilibrium favourable to life and growth" (1971: 127). And

DOI: 10.4324/9781003441311-5

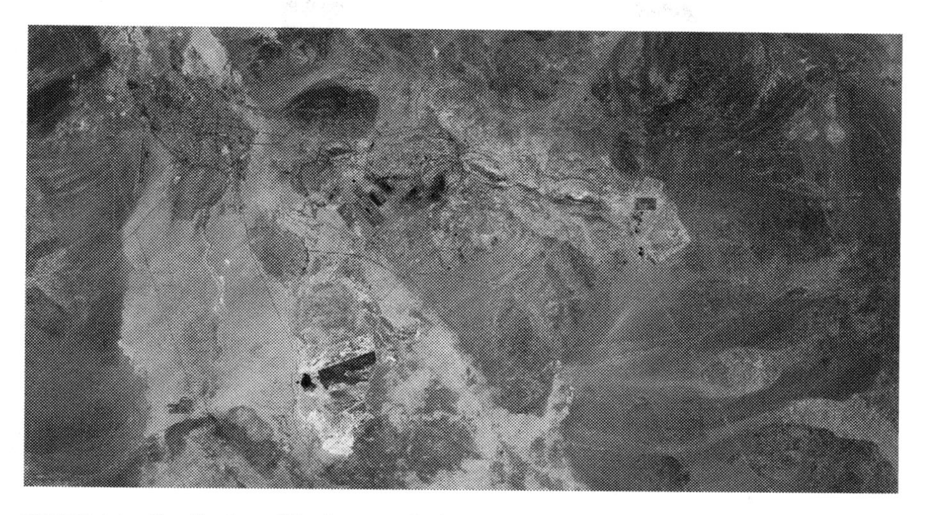

FIGURE 4.1 Sentinel satellite image of Chuquicamata town, mine and tailings.

Source: Copernicus Sentinel data (2023), processed by ESA, CC BY-SA 3.0 IGO.

in the same year, Howard Odum asked: "How much longer will the biological cycles in the uninhabited environments be able to absorb and regenerate the wastes and thus prevent self-poisoning of waters and atmosphere?" (1971: 7). Odum is known as a pioneering ecologist, but as Tega Brain points out, "the history of ecology is enmeshed with systems theory and presupposes that species entanglements are *operational*" (2018: 153). This operative framing of environment is evident in Odum's presupposition that the function of uninhabited regions is to absorb and regenerate human and industrial wastes.

Half a century later, Crary's verdict on the concerns articulated by Mumford and Odum is damning: "Because the global economy no longer has any long-term prospects", he writes, "one, last, mad spree of plunder is now ongoing all over the planet" (2022: 27). Reading the current commentary surrounding energy transition, it is hard not to connect this *mad spree of plunder* described by Crary to the oft-repeated statistics regarding the resource requirements of the green economy. Guillame Pitron, for example, remarks that in the next three decades, humans will consume more metals and minerals than we have in the last 70,000 years (Glenny 2022), while geologist Lucy Crane says that building clean energy infrastructure will require the production of 550 million tonnes of new copper: the same quantity as was mined in the last 5,000 years (2020). The clear message is that the cessation of fossil fuel extraction, so essential to avoid climate catastrophe, will require a massive expansion of mineral extraction. The construction of a renewable energy infrastructure is predicated on a sudden burst of rampant mining activity.

For Mumford, the mine has historically served as the testing ground of numerous mechanical inventions and social practices that were later implemented widely throughout urban society. He lists: "the railroad, the mechanical lift, the underground tunnel, along with artificial lighting and ventilation ... the eight hour day and twenty-four hour triple shift" (1971: 147) as all originating in the mine. Furthermore, Mumford argues, it was mining that

> originally set the pattern for later modes of mechanisation by its callous disregard for human factors, by its indifference to the pollution and destruction of the neighbouring environment ... and above all by its topographic and mental isolation from the organic world.
>
> *(1971: 147)*

More recently, Anna Tsing has proposed a similar argument based on the model of sugarcane plantations, insisting that "factories built plantation-style alienation into their plans" (2017: 40). And the similarity between mines and plantations doesn't end at their labour practices. For the vast majority of human history, mines have operated with a largely plantational logic, extracting a single metal commodity from a specific orebody, echoing the monoculture of the plantation. It is only relatively recently, with growing markets for by-product metals, that mining has begun to diversify the commodities produced at each site, and even then, progress in this respect is still incremental. My focus on mining here follows Mumford's assertion of the mine—and, to a lesser extent, Tsing's of the plantation—as bellwethers of future industrial practices to ask: What might the consequences be of this pattern perpetuating itself in the present?

In *Planetary Mine*, Martín Arboleda writes extensively about recent, and likely future, developments in the mining industry resulting from "the synergistic effect of innovations in robotics, biotechnology, artificial intelligence, and geospatial information systems" whose confluence, he argues, have "exerted a fundamental overhaul in the extraction and processing of minerals" (2020: 11). The consequence of these technological developments has been a consumptive surge enabled by automation and its concomitant exclusion of the human worker, and reduction in labour costs. In Mumford's terms, this exclusion of human labour was a fundamental premise of what he called the *mechanical world picture*, whose origins he perceived as being in astronomy:

> The new world that astronomy and mechanics opened up was based upon a dogmatic premise that excluded from the outset not only the presence of men but the phenomena of life. On this new assumption the cosmos itself was primarily a mechanical system capable of being fully understood solely by reference to a mechanical model. Not man but the machine became the central feature.
>
> *(1971: 33)*

He goes on to draw an explicit connection to industrialisation, stating "what began in the astronomical observatory finally ended in our day in the computer-controlled and automatically operated factory" (1971: 65). And indeed today, as Shannon Mattern has written (2021), the universalising model to which all natural and man-made systems are analogised is less mechanical than it is computational. However, the computational world picture is not only a model or an analogy but can also be seen as materialising literally in the images streamed with mechanistic regularity by fleets of orbiting satellites. The world has become understood not just computationally but also *as a picture*. The inversion of astronomy into earth observation has produced a visual epistemics in which terrestrial conditions are demystified through the application of numerous 'spectral indexes': chromatic calculations that render variables such as vegetative health, mineral composition, hydration, and deforestation as pictorial functions (Cornford 2023). The implication of these imaging practices is not only that planetary health can be diagnosed pictorially, but that fixing the picture equates to fixing the planet. In the epistemics of remote sensing, landscape and image have become conflated with one another.

It is between these poles that the argument below unfolds: between the orbital images of earth observation and the underground extraction of its minerals, between the tailing ponds of northern Chile and the automated mines of the future: between operative and collateral landscapes.

Operative landscapes are spaces instrumentalised for industrial and agricultural production, logistics, manufacture, and so on. They are estimated to account for 14.6% of Earth's surface, or 50% of its land. Their history is surely as long as that of civilisation, and the development and character of various human societies is evident in how they have operationalised the landscapes they inhabit. The term has been taken up recently in landscape architecture, where it has been used to gesture to the spread of urban logistics beyond the city and into the hinterland (Brenner & Katsikis 2020), or in the very different context of community spaces with a shared sense of ownership (North 2012). But my interpretation here is considerably narrower and focused solely on the potential for landscapes to become automated within wider technological metabolisms as a consequence of being imaged. I use the term operative then in the sense of Harun Farocki's *operative image*, where an image functions not representationally but as a stage of a machinic process. In thinking of operative landscapes, I am interested in how the operativity of an image becomes transferred onto its subject. As with the example of remote sensing above, when operative images monitor agriculture and infrastructure, do the landscapes they survey not also become operative within machinic systems? Jussi Parikka concisely describes this transference of operativity from medium to subject, saying: "they look like geographic territories, yet they act like digital data operations" (2023: 170). In such contexts, the operative image becomes a lever that goes beyond replacing the eye in existing machinic processes—as was the case in Farocki's earliest videographic examples—to impose a machinic paradigm on its contents, integrating them in the mechanistic, computational, and commoditising logics of the assembly line or supply chain.

Collateral landscapes, on the other hand, are the inevitable consequence of industrial production. In the examples of mining discussed here, they are primarily tailings ponds and slag heaps, but conceived broadly could include landfill sites, disused quarries, e-waste dumps, polluted waterways, the atmosphere, even the flora, fauna and bodies of workers at these sites. More often referred to as 'sacrifice zones', these often irrevocably damaged environments are the toxic legacy of industrial operations. Tsing has used the term "blasted landscapes" (2014) to discuss similar spaces, but whereas her focus was on landscapes of overt catastrophe such as Gulf of Mexico following the Deepwater Horizon blowout, my use of the term collateral here invokes mundane landscapes that are routinely laid waste by the industrial activities of extraction, refinement, manufacture and disposal. Eyal Weizman articulates this distinction clearly: "collateral effects are structural rather than accidental" (2017: 3). Whereas the blasted landscapes discussed by Tsing rightly generate considerable media attention and public outcry; collateral landscapes, often vast in scale—think, for example, of the Tar Sands in Alberta, Canada—are part of business-as-usual operations that have come to look on their own collateral as an 'externality' whose remedial ecological impact doesn't factor in their profitability calculus. For a truly circular economy to be realised, the very concept of an externality is one of the few things that must be discarded.

Here though, I adopt the word collateral for the benefit of its double meaning. As an adjective, collateral denotes the unintentional consequence of an action, as in the military euphemism of collateral damage. But as an economistic noun, collateral is an asset or security, the inherent value of something owned which can be used as purchasing power. It's worth noting that the (to my mind, highly dubious) economic practices implemented to mitigate climate change have so far relied on exactly this principle of leveraging collaterals against emissions, as with carbon offsetting where, for example, a forest's existing capacity to sequester carbon dioxide is traded as emissions quotas to industry. It is in this double meaning of collateral that the argument of this chapter is contained. Waste is collateral, in both senses of the term. Waste, tailings, and slag are all collateral by-products of refining ores to their metallic elements, but they can also be conceived as collateral, as a resource to be recuperated. If the proposed circularity of the digital economy is to become more than a hypothetical model this collateral needs to be operationalised. So, following Mumford's assertion of the mine as a testing ground of future technics, we can ask: if we are genuinely on the cusp of shifting from a linear to a circular economy, from a carbon combustion economy to a sustainable model of renewable energy and material recuperation, then might we not already be able to see signs of this future emerging in mining practices?

Operative landscapes

In Harun Farocki's *Images of the World and the Inscription of War* (1988), we see a photographic interpreter being trained to recognise the infrared heat signals around parked aircraft to discern how long it is since they were flown, or even how long it was since an empty space was inhabited. When Farocki spoke of the operative image, he was not initially referencing this type of visual labour, but pointing to the insertion of images into already machinic processes, where they either replaced human oversight, provided remote scrutiny, aided automation, or served as proof (sometimes propaganda) of machinic efficiency. In these instances, the image had become one stage in an automated process, but in the 35 years between the completion of Farocki's film and now, the operative image has also encroached upon the labour of the photographic interpreter. The training documented in this scene is fast being outsourced to dataset training of machine interpretation, extending the automation from physical labour to inspectional labour, as Farocki later wrote "industrial production abolished manual work – and also visual work" (2001).

Codelco, the Chilean state-owned company which runs the mine at Chuquicamata, has not invested in automation technologies to the same extent as its wealthier multinational competitors, but elsewhere—particularly in Western Australia—the story is somewhat different. Rio Tinto's 'Mine of the Future' project for example, launched in 2008 and now largely implemented at their Pilbara site in Western Australia, boasts of automation being employed throughout the extraction workflow. Autonomous drill rigs are operated remotely from 1,500 km away in Perth, with one employee simultaneously controlling three rigs. Here, the operative image has abolished the manual labour at the mine and replaced it with remote visual labour. Throughout this mineral supply chain, cameras and other sensors perceive the environment, relaying images or data visualisations aggregated from multiple apparatuses to their remote monitors, or simply triggering the next machinic sequence. Driverless tipper trucks run day and night at Gudai-Darri, refuelled "using a vision sensing and direction system to locate the position of the truck's fuel tank and connect it to the fuel nozzle" (Rio Tinto 2022). This automation at the mine itself is then extended into their AutoHaul rail network. Throughout this chain, the image inhabits exactly the limited role originally described by Farocki, functioning to displace manual and visual work, and therefore decrease labour costs.

But, as the refined metal moves beyond a single site, or leaves one corporation's control software, it enters as data into another network of operativity: that of logistics and shipping. In *The Ruin, the Jewel and the Chain*, Alejandro Donaire Palma discusses the integration of Chilean copper mining with international exports from the port of Valparaiso, where software developed by Spanish IT company Indra is used to systematise the logistics chain. The software in question, SILGOPORT, "position[s] and map[s] the physical and documentary flow of freight … integrating different operators in interdependent processes" resulting in what Palma refers to as a "common territorial diagram" (2021: 22). The space between mine and port is mapped, not as a terrain or an environment but as a flow of commodities. The landscape is rendered first as a repository of resources and then as channel of communication for their export. And the software's ownership demonstrates the persistence of colonial power structures in determining the operative priorities with which those landscapes are viewed. The supply chain integration discussed by Palma is now also proposed to extend across the ocean, where, as Miriam Posner writes, "technologists in the shipping industry envision a near future in which one captain controls a fleet of crewless ships" (2022) whose position and conditions would be monitored by satellite imaging and geolocational data.

A similar scenario of the operative image inflecting future agricultural practices could also be sketched with respect to Tsing's model of the plantation. Since monitoring of farmland has become one of the most common commercial applications of satellite imagery, connecting an infrared, daily pictorial analysis of soil moisture or leaf chlorophyll to an automated irrigation system seems a plausible and logical step for a high yield farm. To do so turns the landscape into a function of the image, optimising plant reflectance to correspond to a chromatic diagnosis of health. Farming has operationalised landscapes for centuries, but continual feedback from orbital imaging has the potential to transform the paradigm of this operativity to something in which human decision-making takes a back seat. In these two examples, the future potential to integrate the entire logistics chain for mineral and vegetal commodities is clear. From cultivation or drilling, through harvest and export, managing supply to meet fluctuations in demand or vessel capacity can be monitored and automated through the integration of operative imaging technologies.

But image operations do not only enable extraction and logistics, as Martín Arboleda explains, they also precede mining operations: "the introduction of geospatial information systems to mineral forecasting and geological surveying, has allowed engineers to produce highly accurate representations of the subsurface, making the extraction of low grade ore bodies profitable for the first time in history" (2020: 4). In contemporary geophysical prospecting, these mineral forecasts are produced by combining and visualising data from multiple sources. Drone-mounted aerial surveys record multispectral photographs, producing a high resolution topographic mosaic of the site whose variable reflectance can be analysed to determine the composition of any exposed rock. Colourised models of drill-core samples are superimposed on this landscape, producing vertical profiles which can be stitched together with data from ground-penetrating radar to map the strata beneath. Three-dimensional visualisation software aggregates these multiple sources, enabling the volume of target minerals to be estimated. This complex and often protracted sequence of technical procedures applies a variety of data visualisation techniques to the landscape, and throughout images perform pivotal operative roles. These acts of data extraction—whether rendering areas of ground as chromatic values or counting microsecond variations in the subterranean reflection of an impulse—all rely upon a numerical calculus that depicts the landscape as a series of coded values. Prospecting thereby extends the quantitative condition of digitality onto its subject: colour values denote mineral deposits whose volumetric values produce monetary value. These abstractions render the landscape itself as a purely operative resource whose social, cultural, and ecological value is completely elided from, and obfuscated by, these machinic processes. In short, the digitisation of images hastens the quantification of that which lies in the frame. These computed images of the world, produced by the predictability of gravitational orbits that Mumford sites as the origin of the mechanical world picture, computationally model landscapes and planetary ecosystems.

The material and semantic consistency between the circuits and signals of the image and those of computation not only produces a world understood pictorially but also gives images a potentially decisive capacity. At present, the prospecting tasks described above are still performed by technicians, but the quantity of earth observation data currently produced begs the question of how long such an approach is practicable. In 1973, NASA carried out one of the earliest multispectral experiments, shooting continuously from Skylab using six synchronised cameras, each loaded with different 70 mm film stocks and fitted with different filters. Project scientist Kenneth Demel, interviewed on television, said that to photograph the same area by aircraft would produce "an almost unmanageable collection of photographs" (1973), yet now, 50 years later, a similar quantity is transmitted daily from satellites. The sheer weight of this data necessitates computational aggregation and analysis. Digitisation has made the management of such vast image archives routine, and the current, much-hyped, advances in artificial intelligence are making machine interpretation of such image archives feasible. We may still be some years away from automated geophysical prospecting, but for many other data analytics industries, the term "actionable insights"—meaning computational conclusions drawn from a dataset—denotes the now commonplace inclusion of automated analysis in the decision-making loop. This too is a consequence foreseen and critiqued by Mumford, who wrote: "automation has a qualitative defect that springs directly from its quantitative accomplishments: briefly it increases probability and decreases possibility" (1971: 185). By its very nature, the machine aggregation of massive datasets leads to conservative conclusions that hug the mean. Tega Brain offers a particularly poignant example of this, in which the rapid depletion of atmospheric ozone was initially discarded as an outlier by NASA's data processing (2018: 156). Automated decision-making then is incapable of making paradigm leaps or of dramatically altering course, which—ironically—is exactly what is urgently required of energy and industrial policy.

From the use of spectral indexes in mineral forecasting, through the operative systems automating its subsequent extraction, to its logistical flow to and between shores, the future vision projected by current mining practices is one in which indexes of commercial value are tantalisingly close to being integrated with indexed global mineral reserves and indexed forecasts of mineral supply. We are moving from the mechanical world picture to a pictorially computed planet. Throughout this accelerationist's dream of data-driven extractive flows, investment is focused on expanding and intensifying automation at the beginning of the supply chain. If, as Mumford argues it once was, mining remains today a bellwether industry of our future industrial trajectory, then it is hard to see how its present practices point to anything other than an intensification and automation of extraction. So, let us now turn to its obverse overburden, to the displaced wastes and discarded detritus produced not just at the end of the supply chain, but throughout its various operations.

Collateral landscapes

In Chuquicamata, the health of mine labourers and their families is not the only motivation for their relocation. The evacuation of the town will also bring tangible economic benefits to Codelco. As the ore quality in the main pit gradually declines, the proportion of collateral to copper gradually increases, and the mine must expand. Once the houses of Chuquicamata lie empty, they will become engulfed in the operation of the mine, buried beneath its expansive spoil heaps. At this point, the housing of ex-workers will become indistinguishable from the overburden they themselves hauled: the rubble of homes sedimenting beneath the rubble of unprofitable spoil. The mining town, the lives and community of its inhabitants, and the slag heap itself will all become collateral landscapes, folded together in a single site of valueless abandon.

The landscape surrounding Chuquicamata is so arid that the water used in processing the ore is recycled repeatedly by being pumped into tailings ponds after use, left to stand until the toxic sludge has settled to the bottom, then pumped back into the mine for reuse. This gravitational sedimentary process echoes the natural formation of mineral deposits in magma chambers by fractional crystallisation, where heavier minerals are deposited at the bottom of pooling magma. But, unlike igneous deposits of chromite and ilmenite, which are mined for their chrome and titanium, the sediment settling in tailings at Chuquicamata—whose weight belies its constituent heavy metals—is regarded as a waste material.

Tailings ponds and slag heaps are synonymous with mining, the inevitable collateral of refining ores into primary commodities. The realisation of a circular economy is currently predicated on a radical expansion of mining, but the collateral generated by this expansion remains a so-called externality, and to assume that its volume will increase in direct proportion with the quantity of metal mined is naive. As Arboleda points out: "the economic profitability brought about by the smart and robotised mine, pales in comparison to its material footprint, as an average large-scale extraction site produces up to 1000 times more solid waste than those working with older technologies" (2020: 11). The consequence of mining's multiple automations has been to "radically upscale its metabolic processes" (2020: 10), enabling the profitable processing of far lower grades of ore. While the collateral damage to the bodies of mine workers are doubtless reduced by automation, its collateral landscapes proliferate dramatically. The inevitable result of processing lower-grade ores is a disproportionate increase in both the quantity of waste and the energetic cost of refinement.

The finely pulverised minerals known as tailings are the most reactive rocks on Earth and they are also abundant, with current estimates of global production between 10 and 14 billion tonnes annually. Yet evidence is fast accruing that many traditional waste materials of major ore extraction are rich in secondary commodities that are suddenly highly valuable due the digital economy's hunger for by-product metals and trace elements. Geologist Eimear Deady, for example, has shown that Bauxite wastes, the red mud residue of aluminium refinement, can be reused for the production of rare earth elements (REE) and other critical metals, such as gallium (2020). And a similar study on the tailings of a closed tin mine in Indonesia has also demonstrated the presence of "abundant REE bearing minerals" (Deon et al 2021). But it's not only through reprocessing existing tailings that mining could circularly operationalise its existing collateral. The Mount Keith mine in Western Australia extracts nickel from ultramafic rocks, pumping the tailings into a pond almost 5 km wide. Here, Sasha Wilson found that the waste material carbonates naturally, sequestering atmospheric CO_2 in the process. Although at present only 1% of the tailings material carbonates, this represents an 11% offset of the mine's carbon emissions. If this could be increased to just 10% the mine would become a net zero emitter, and, if increased further, it could be operationalised as a negative emissions technology (2021). What is fascinating about Wilson's talk though is how she couches this untapped potential for carbon sequestration in economistic terms: as a "service" that could presumably be traded as an emissions offset.

Adopting an economic lens leaves the opportunity this finding represents hostage to the profit motive. Economics confuses, even conflates, viability with profitability, preventing the implementation of circular modes of production, as is currently seen in post-consumer waste streams. The conductive coatings of touchscreens, for example, are made of indium tin oxide, a compound that uniquely combines the properties of transparence and conduction. Indium is a by-product of the refinement of zinc from sphalerite deposits. At present, only 1% of global supply is currently recycled from end-of-life devices, because, for this process to become industrially viable—meaning profitable—the price of raw indium would need to exceed $700/kg. Without some multilateral consensus on the prioritisation of resource consumption for energy transition, as Drew Pendergrass and Troy Vettesse point out, "markets decide whether the steel goes into wind turbines or luxury SUVs" (2022: 45). This absurd situation, in which valuable materials are discarded and profitability defines priority, is perpetuated by the failure to account for the collaterals, the so-called externalities that can be discarded with minimal cost to industry, but long-term costs to the environment. In her book on rare earth landscapes, Julie Klinger tackles this directly, stating:

> There are no 'externalities' in geography: the very word reflects a way of thinking that does not match reality. As residents in an integrated biophysical Earth

> system, there is no part of the Earth that is external to our affairs. Pollutants do not respect boundaries, nor do our efforts to acquire the elements essential to contemporary life.
>
> *(2017: 8)*

These collaterals, the structurally neglected negative effects of industrial extraction, are not externalities but raw materials. The logic of thinking their carbon sequestration capacity as an economic service means that their current neglect and negative environmental impacts should be considered costs. If we are to place these materials back inside an economic frame, then perhaps we should also think them in semantically economic terms: shifting them from collateral effects to collateral assets. For a circular economy to be realised, this collateral must become economically internalised; it must be treated as a resource. Again, this thinking is not new. Odum observed that "the high concentration of chemical substance in [human] wastes are actually rich reserves of energy" (1971: 291) (a proposition which has been speculatively realised in artist Martin Howse's recent *Tiny Mining* project (2020)). For Tsing, capitalism excels at what she calls *salvage accumulation*, the co-opting of "living things made with ecological processes … for the concentration of wealth" (2017: 63). Yet, to date, the capacity of industrial processes to salvage their own collateral remains pitiful. Throughout the supply chain, unsalvaged externalities proliferate as unrecuperated energy, unrecovered metals, unsequestered emissions, and unrecycled products.

According to Kate Raworth, this could be addressed through a "switch from taxing labour to taxing non-renewable resources" (2020: 238), a policy that gets directly to the heart of the opposition established here between operative and collateral landscapes. Raworth argues that in a context where more than 50% of the EU's tax revenue comes from labour, it is "no surprise that industry's response has been to focus on increasing labour productivity by replacing as many workers as possible with automatons" (2020: 238). Shifting the burden of taxation from employees to resources could shift the material and financial investment currently expended in automating exploration and extraction of novel resources to one that prioritises energy efficiency, and maximises the exploitation of its own collateral. The research currently underway into the recycling of these collaterals—be they from mining or end-of-life digital devices—is taking place predominantly in academic contexts, and at far from the necessary scale to make a significant contribution to planetary industrial metabolisms. That investments in automation far outstrip those directed towards recuperating and remediating by-products, tailings, discards, and obsolete materials is a consequence of an economic policy which perpetuates the fallacy of externalities.

Towards an ouroborian economy

For her project *The Iron Ring*, artist Cecilia Jonsson harvested 24 kg of Imperata Cylindrica, an invasive weed also known as blood grass due to its scarlet tips, from

FIGURE 4.2 Cecilia Jonsson, *The Iron Ring* (2013). Documentation from the harvesting process.

Source: Photo by Linda Tulldahl.

the banks of the Rio Tinto (Figure 4.2). The river owes its name and red hue to the quantity of iron dissolved in its waters, which flow from mountains that have been mined for their precious metal deposits for 5000 years. This dissolved iron is absorbed into the roots, stems and leaves of the grass, which, unlike many other plant species, proliferates in the acidic soil of the river banks. Having harvested the grass, Jonsson proceeded to extract its iron content: cleaning, drying and incinerating it before smelting the iron from its ash. From the 24 kg of grass, she eventually produced 2 g of iron which was cast into a single ring, whose circular shape reinforces the circularity of her process. Through this method, Jonsson proposes "a scenario for iron mining that, instead of furthering destruction, could actually contribute to the environmental rehabilitation of abandoned metal mines" (2014).

Imperata Cylindrica is what is known as a hyperaccumulator, a plant that thrives in soil whose metal content is toxic to most other flora by absorbing those metals. James Bridle has written about current research into the commercial application of this process—known as agromining—in Northern Greece, where three plants are being farmed for their ability to absorb nickel, an increasingly common metal in the cathodes of lithium-ion batteries. These plants, he observes, were considered weeds and pests "until we realised that they were aligned with our own needs" (2022: 310). Their hyperaccumulating potential has suddenly endowed them with an operative function. And this shift in perspective, from weeds to crops, could be extended to the pollutants, effluents, and other collaterals discussed above. As

Bridle writes, such processes will never entirely replace the need to mine new materials. But, given their potential to both provide a secondary source of metals critical to energy transition and their capacity to reduce and remediate the environmental degradation of mining, such process has the potential to salvage valuable materials from waste and restore ecological balance in collateral landscapes.

In 1971, when both Lewis Mumford and Howard Odum questioned the planetary capacity to remediate its industrial devastation, Earth Overshoot Day —the date when humanity's annual demand for ecological resources exceeds what Earth can regenerate—was estimated to have been on December 25th. Fifty years later, it was on August 3rd. To achieve the astounding expansion of mineral extraction, that is routinely quoted as necessary for energy transition, without rapidly decreasing consumption in other areas, will inevitably force this date even further beyond planetary boundaries. For circularity to be realised, the operative eye must be cast upon the vast volumes of collateral generated throughout the digital economy. Waste is a resource. Obsolescence is a choice. The wastage integral to extraction must become integrated into production. The snake must consume its own tailings.

References

Arboleda, M. (2020) *Planetary Mine: Territories of Extraction under Late Capitalism*. London: Verso.

Brain, T. (2018) "The Environment Is Not A System", *Advanced Research Projects Agency* 7, no. 1, 153–165.

Brenner, N. & Katsikis, N. (2020) "Operational Landscapes: Hinterlands of the Capitalocene", *Architectural Design* 90, no. 1, 22–31.

Bridle, J. (2022) *Ways of Being: Beyond Human Intelligence*. London: Penguin.

Cornford, S. (2023) *Spectral Index*. Melbourne: Avantwhatever. Available here: https://avantwhatever.xyz/w/tsSQUPENFwnYfwJTche3jh, (accessed 11/10/2023).

Crane, L. (2020) "Mining Our Way to a Low Carbon Future", January 21 2020, video, 13:36. Available here: https://www.youtube.com/watch?v=aWTkiQ64u_U, (accessed 11/10/2023)

Crary, J. (2022) *Scorched Earth: Beyond the Digital Age to a Post-Capitalist World*. London: Verso.

Deady, E. (2020) "Critical Metals in Bauxite & Red Mud. Under Mined Resources?", July 8 2020, video, 1:16:41. Available here: https://www.youtube.com/watch?v=eq9vha2AUMg, (accessed 11/10/2023).

Demel, K. (1973) "Skylab Experiment S190 Multispectral Facility", June 30 1973, video, 3:11. Available here: https://www.youtube.com/watch?v=07o1JrH8l28, (accessed 11/10/2023).

Deon et al. (2021) "Rare Earth Elements (REEs) in Mine Waste: A Way to Solve the Rising Worldwide REEs Demand?". Available here: https://research.utwente.nl/en/publications/rare-earth-elements-rees-in-mine-waste-a-way-to-solve-the-rising-, (accessed 11/10/2023).

Images of the World and the Inscription of War (1988). Harun Farocki [video] Berlin: Harun Farocki Film Production.

Eye/Machine (2001). Harun Farocki [video], Berlin: Harun Farocki Film Production.

Glenny, M. (2022) "The Scramble for Rare Earths", BBC Radio 4, September 26 2022.

Howse, M. (2020), *Tiny Mining*. Available here: https://v2.nl/works/tiny-mining, (accessed 11/10/2023).

Jonsson, C. (2014) *The Iron Ring*. Rotterdam: V2_Publishing.
Klinger, J. (2017) *Rare Earth Frontiers: From Terrestrial Subsoils to Lunar Landscapes*. Ithaca: Cornell University Press.
Mattern, S. (2021) *A City Is Not a Computer: Other Urban Intelligences*. Oxford: University of Princetown Press.
Mumford, L. (1971) *The Pentagon of Power*. London: Seckler & Warburg.
North, A. (2012) *Operative Landscapes: Building Communities Through Public Space*. Basel: Birkhäuser.
Odum, H. (1971) *Environment, Power and Society*. London: Wiley-Interscience.
Palma, A. D. (2021) "The Ruin, the Jewel and the Chain", in Neilson, B. & Rossiter, N. (eds) *Logistical Worlds: Infrastructure, Software, Labour, No. 3 Valparaíso*. London: Open Humanities Press.
Parikka, J. (2023) *Operational Images: From the Visual to the Invisual*. Minneapolis: University of Minnesota Press.
Pendergrass, D. & Vettesse, T. (2022) *Half Earth Socialism: A Plan to Save the Future from Extinction, Climate Change and Pandemics*. London: Verso.
Posner, M. (2022) "Ghost Ships", *Logic(s)* 18, December 21 2022. Available here: https://logicmag.io/pivot/ghost-ships/, (accessed 11/10/2023).
Raworth, K. (2020) *Doughnut Economics: Seven Ways to Think Like a 21st Century Economist*. London: Random House.
Rio Tinto (2022) "Rio Tinto: Look Inside a Mine of the Future", last updated August 12 2022. Available here: https://www.riotinto.com/news/stories/look-inside-future-mine, (accessed 11/10/2023).
Tsing, A. (2014) "Blasted Landscapes (and the Gentle Arts of Mushroom Picking)", in *The Multispecies Salon*, E. Kirksey (ed.), Chicago: Duke University Press, pp. 87–109.
Tsing, A. (2017) *The Mushroom at the End of the World: On the Possibility of Life in Postcapitalist Ruins*. Princeton: Princeton University Press.
Weizman, E. (2017) *The Least of All Possible Evils: A Short History of Humanitarian Violence*. London: Verso.
Wilson, S. (2021) "Waste Not, Want Not: Using Waste Minerals as a Resource for Carbon Sequestration and Metal Recovery", August 11 2021, video, 46:32. Available here: https://www.youtube.com/watch?v=tTw5gK39RFY, (accessed 11/10/2023).

5

FRAMING THE (UN)SUSTAINABILITY OF AI

Environmental, social, and democratic aspects

Irene Niet, Mignon Hagemeijer, Anne Marte Gardenier, and Rinie van Est

Introduction

A growing number of policymakers, businesses, and academics perceive digital technologies as crucial for reaching sustainability goals (European Commission 2021). The European Commission is currently (in 2023) focussing on finding synergies between the green transition and digitalization, for "a sustainable, fair, and competitive future" (European Commission 2022a). In particular, artificial intelligence (AI) is perceived as a core technology in these synergies (European Commission 2022a; European Commission 2022b). We broadly conceptualize AI as algorithms with the capacity to gather and/or analyse large amounts of data; produce reports and/or act autonomously on the basis of that analysis; learn from the results of its actions independently from human intervention; and/or adapt itself to incorporate learning (Kalogirou 2007; Poole and Mackworth 2010; Sarangi and Sharma 2019). While there is general optimism about AI supporting sustainability, tales of caution are increasing (Crawford 2021). Between 2018 and 2023, the growing political attention at the EU level for the twin transition (European Commission 2022c) intensified the academic debate regarding AI. In this chapter, we show the different discussions in academia surrounding AI (un)sustainability during the period.

To gain a comprehensive understanding of the academic debate surrounding AI (un)sustainability, we use two overarching dimensions: the ecological sustainability of AI, and the social or democratic sustainability of AI. The ecological sustainability of AI concerns its impact on the natural environment (Cherubini et al. 2019; Nilashi et al. 2019). As there is a substantial literature regarding the possible positive impacts of AI for the environment (Asha et al. 2022; Kaack et al. 2022), we conducted a literature review to understand the *negative impacts* of AI for sustainability. Using electronic databases (Jstor, arXiv, Google Scholar, and Philpapers), we identified English language publications (including preprints, review articles, reports, and original articles) through combining search terms for AI

DOI: 10.4324/9781003441311-6

("artificial intelligence", "machine learning" and "deep learning") with the terms "climate change", "climate impact", "environmental impact", "resources", "water use", "energy use" or "sustainability". We excluded publications which solely discussed applications (including robotics, self-driving cars, or other AI-related fields) or benefits of AI. To minimize omissions, we conducted an initial search at the end of 2022, followed by an updated search in August 2023. In total, 40 publications were included in the review.

The second overarching dimension is the social or democratic sustainability of AI. This refers to the impact of AI on equality and democratic processes in society (Elzen and Wieczorek 2005; Sovacool et al. 2017). We conducted a second literature review, using the Scopus database, as it contains more social science publications, fitting with the focus on social impact. We included English language (review) publications, and used the terms "artificial intelligence", "machine learning", and "deep learning" in combination with the terms democra*, or just*. Using the PRISMA method (Page et al. 2021), we selected the publications which were both available and relevant, and excluded publications with insufficient focus on justice or democracy issues, or strong emphasis on specific applications of AI. We included 33 publications in the review.

In the section 'AI as a technology', we discuss the technological workings of AI. This is the basis for the subsequent sections 'Ecological sustainability' and 'Social and democratic sustainability', in which we present our findings from the two literature reviews regarding AI's ecological and socio-democratic (un)sustainability. We then discuss the coping strategies identified in the review to deal with AI's impact. In the final section, we discuss the interdependence between the effects of AI on both ecological and social or democratic sustainability.

AI as a technology

To grasp the sustainability of AI development and implementation, the full spectrum of environmental, social, and democratic costs of AI should be considered. For this, we adjusted the Greenhouse Gas Protocol's framework (Greenhouse Gas Protocol 2013) to fit the AI ecosystem, shown in Figure 5.1. In the literature, five key themes of the AI ecosystem can be discerned: (1) AI development and (2) AI in use, which are both part of the AI lifecycle, and (3) raw materials and processing, (4) core hardware, and (5) e-waste, which are part of the supporting infrastructure. An overview is given in Table 5.1.

AI lifecycle

AI development

This cycle starts with AI development. To develop an AI model, data needs to be collected and processed, (hyper)parameters are tuned and the main algorithm is developed (Brevini 2020; Dacrema et al. 2019). In the development phase,

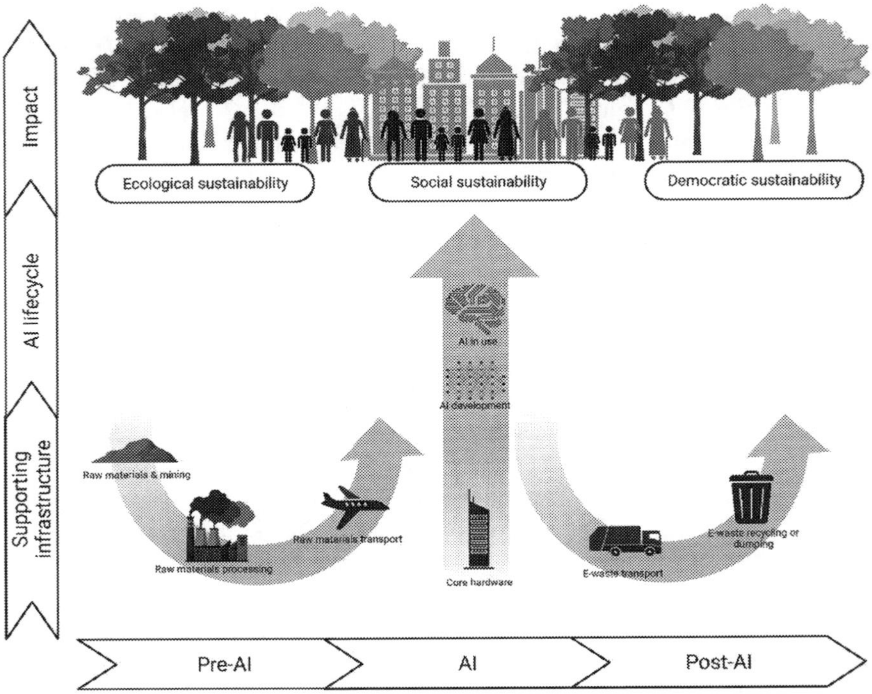

FIGURE 5.1 AI ecosystem and its impact. Based on Greenhouse Gas Protocol (2013).

decisions are made, such as the desired accuracy of the model, its training time and its method of (automated) learning, which impacts the functioning of the model (Badar et al. 2021; Bender et al. 2021; Stray et al. 2021; Wu et al. 2022). This functioning is assessed during the final test and verification stage of the AI development, where it becomes clear whether the AI needs to be further developed and retrained or is ready to be put into use (Strubell et al. 2020).

AI in use

After the AI model is developed, it is put into practice. During this phase, the AI model is used simultaneously by a multitude of users, often repetitively (Schwartz et al. 2020). The AI model has to generate output, as well as learn and retrain based on different stimuli, such as new data and the reactions to the decisions made by the AI (Kar et al. 2022).

Supporting infrastructure

Some authors argue that besides the AI lifecycle, AI requires supporting infrastructure (Clutton-Brock et al. 2021). This infrastructure consists of three elements: raw materials and processing, core hardware, and e-waste.

TABLE 5.1 The AI ecosystem and its impact on ecological, social, and democratic sustainability

Core elements of AI ecosystem	*Technological components and processes*	*Aspects of ecological sustainability*	*Aspects of social and democratic sustainability*	*Coping strategies*
AI lifecycle				
Development of AI model	• Data collection • Data pre-processing • (Hyper)parameter tuning • Developing main algorithm o Size o Training o Type of algorithm • Testing, verification, and validation	• Energy usage of data collection and (pre-) processing, (hyper)parameter tuning, training of the model, testing, and verification of the model • Energy usage depends on the size of the model and the type of (learning of the) model	• Increased inequality due to biases and systemic problems embedded in the AI model • Limited training • Limited transparency and no democratic control	• Developing a clear, standard measure for ecological impact • Developing simpler models (fewer parameters) • Limited training (smaller datasets) • Limiting usage, number of users and number of adjustments • Designing-against-discrimination or adjustments • Empower humans to rectify AI decisions
AI in use	• Inference • Usage • Number of users • Monitoring • Implementation	Energy usage depends on: • the efficiency and accuracy of the model • amount of usage • number of users • the complexity of the model	• Decrease in privacy due to continuous sharing of personal data • Decreased human decision-making and autonomy • Inequality as a result of biased models, non-transparency and decreased autonomy • Inequality due to limited benefits of using AI models • Changed access to information	• Practice caution when implementing AI • Policies & requirements for AI corporations • Human rights impact assessments or disclosure schemes • Establish institutions for public oversight and monitoring of AI

(*Continued*)

TABLE 5.1 (Continued)

Core elements of AI ecosystem	*Technological components and processes*	*Aspects of ecological sustainability*	*Aspects of social and democratic sustainability*	*Coping strategies*
Supporting technological infrastructure				
Raw materials & processing	• Raw materials mining • Processing of materials • Transportation of materials & hardware	• Emissions & waste due to o mining o processing of raw materials o transport of processed materials	• Unsafe working conditions • Data as a raw material without compensation	• Developing a clear, standard measure for social and democratic impact • Ensuring software fits with hardware • Datacentres o PUE measure o Carbon offset scheme
Core hardware	• Processor • Datacentres o Server type o Location & space o Energy & water use	• Interoperability and efficiency • Energy and water usage o partly depends on the server type o Physical space usage	No research found	
E-waste	Dysfunctional or obsolete components	• Limited & unsafe recycling • Burning and dumping of waste	Unsafe working conditions	

Source: Authors.

Raw materials and their processing

This phase includes mining raw materials for the hardware for datacentres and the chips for the software to run, as well as materials for the end-user devices (Crawford 2021; Dauvergne 2022; Rohde et al. 2021). These raw materials often contain rare metals (e.g., lithium, ion, coltan, and cobalt), which need to be processed into chips and other hardware (Crawford 2021; Dauvergne 2022; Lannelongue et al. 2021). After processing, the products need to be transported (Crawford 2021).

Core hardware

The core hardware includes the processor used in computers and datacentres to apply the AI. Hardware and AI software are strongly related: certain processors function best with specific AI models (Henderson et al. 2022; Shahid and Mushtaq 2020; Strubell et al. 2020).

Most AI models require datacentres to function. Although theoretically it is possible to develop and run an AI on site, most companies or research centres opt for datacentres, run by service providers, to train, store and run an AI. Datacentres differ in the server type they use, the physical space they occupy, and their energy and water usage to keep the datacentre operational, cool the hardware and light the facility (Bender et al. 2021; Brevini 2020; Lacoste et al. 2019; Lannelongue et al. 2021; Patterson et al. 2021; Schwartz et al. 2020; Strubell et al. 2020).

E-waste

At some point, any hardware component of the digital infrastructure will need to be replaced. Sometimes, this is because components stop working properly, but more often, with the rapid developments of ICT, components become outdated (Brevini 2020; Crawford 2021). These components need to be disposed of.

Ecological sustainability

When it comes to the ecological sustainability of AI and its ecosystem, the literature emphasizes the great complexity in assessing its precise impact (Nordgren 2022). In this section, we follow the five key elements of the AI ecosystem to gain a better understanding of AI's ecological unsustainability, as shown in Table 5.1.

AI lifecycle

AI development

Data collection, data pre-processing, tuning of the (hyper)parameters and development of the main algorithm all use energy (Brevini 2020; Dacrema et al. 2019). It has been estimated that training a natural language-processing AI model produces

300,000 kilograms of CO_2 emissions (Bucknall and Dori-Hacohen 2022). Additionally, decisions regarding the model's desired accuracy, its training time and its method of (automated) learning, can have major impacts on ecological sustainability, as this affects energy usage during development as well as when the model is in use (Badar et al., 2021; Bender et al., 2021; Schwartz et al. 2020; Stray et al. 2021). Furthermore, the cycle of final testing and verification of the model can quickly increase its environmental unsustainability (Strubell et al. 2020). The growing complexity of models increases the ecological unsustainability of AI model development (Gutiérrez et al. 2022; Probst 2022; Wu et al. 2022).

AI in use

Most calculations regarding the ecological sustainability of AI become even more challenging when the AI model is in use. Computer hardware supplier NVIDIA estimated that the ecological impact of training the AI model is only a tenth of the programme's impact while in operation (Patterson et al. 2021). There are ways to adjust AI models to ensure they are more efficient, for example by grouping them in datacentres and incorporating ways the AI models can run on renewable energy (Borowiec et al. 2022). The continuous running of AI models by multiple users complicates taking such 'greening' measures when AI is already operationalized (Henderson et al. 2022; Lacoste et al. 2019). Although a simpler model might be more resource efficient and take less time for calculations, there could be a rebound effect: the simplicity of such models invites more use, as well as more monitoring of efficiency, increasing the ecological costs in the operational phase (Budennyy et al. 2022; Kopka and Grashof 2022; Kum et al. 2022; Lannelongue et al. 2021; Stoll et al. 2022).

Supporting infrastructure

Raw materials and their processing

The extraction of raw materials needed for the AI model's supporting infrastructure often has a negative ecological impact, due to the dumping of waste material in rivers and the production of radioactive waste (Crawford 2021; Robbins and van Wynsberghe 2022). The ecological impact also includes the emissions and waste resulting from the processing of raw materials into chips and other hardware (Dauvergne 2022; Lannelongue et al. 2021). The ecological impact of this processing step is under-researched. Similarly, there is little known about the emissions related to transport of the processed materials (Crawford 2021).

Core hardware

Many processors can only function at maximum efficiency with specific AI models. This means that more efficient AI models may negatively impact the ecological sustainability of the processor if they are run on ill-fitting hardware (Henderson et al. 2022; Kum et al. 2022; Shahid and Mushtaq 2020; Strubell et al. 2020). The ecological impact of datacentres depends on their server type, physical space usage and energy and water usage (Bender et al. 2021; Brevini 2020; Lacoste et al. 2019; Lannelongue et al. 2021; Nost and Colven 2022; Patterson et al. 2021; Schwartz et al. 2020). Opting for datacentres instead of on-site technology is often more ecologically sustainable, as the service provider can group the run time of AI models, and base this, for example, on the availability of renewable energy (Cioca and Schuszter 2022; Gomez et al. 2022; Kar et al. 2022; Kum et al. 2022; Yang et al. 2022).

E-waste

Although digital technologies, including AI models, could help to reduce waste in general (Rosário and Dias 2022), they also produce e-waste. Only one-fifth of the total amount of e-waste is currently being disposed of in a safe manner (Dauvergne 2022). Although recycling of e-waste could reduce the ecological impact of the raw materials phase, this recycling is often done by hand in developing countries, exposing the human pickers to hazardous chemicals (Robbins and van Wynsberghe 2022). Alternatively, if components are not recycled, the e-waste may be burned in open fires, or dumped in landfills, allowing these chemicals to end up in the earth, the groundwater and the air (Samuel and Lucassen 2022). This ecological sustainability impact of e-waste is also rarely quantified.

Social and democratic sustainability

The second literature review concerned the impact of AI on social and democratic sustainability. Here, we again use the framework of the AI ecosystem and its five themes to clarify the impact of AI models and their usage. We show our results in Table 5.1.

AI lifecycle

AI development

Concerning specifically the AI lifecycle, decisions made during AI development, and the way the AI model is used and implemented, greatly affect its social and democratic sustainability. The first major impact identified in the review is the integration of biases and systemic problems in AI models. Systems are not flawless; they can make mistakes due to poor design or coding, and could embed social prejudices and power structures in such a way that these become 'invisible' (Bannister and Connolly 2020). This is because the AI models are trained using historical data, which often contains such biases (Starke et al. 2022). Based on flawed data, these systems then make assumptions regarding people's personal preferences and future behaviour, which could be wrongful and lead to justice issues (Halsband 2022; Zimmermann and Lee-Stronach 2022; Završnik 2020).

The second major social impact stemming from the development of an AI model is a limitation to the training of AI models. AI models cannot really be trained to handle unforeseen circumstances, and they lack human characteristics such as sympathy, empathy, insight, and experience, which humans use to deal with such situations. As a result, all occurring situations are converted into codes and numeric ranking systems, lacking a 'human perspective' to deal with unforeseen circumstances (Bannister and Connolly 2020; Sadowski et al. 2021).

This is connected to the third major social and democratic sustainability issue of AI development: its limited transparency. Private companies are not always required to be transparent about the model, as it is considered a trade secret (Martínez-Ramil 2021). In a democracy, this can be problematic: these AI models cannot be subjected to public scrutiny (Buhmann and Fieseler 2021). This can lead to issues regarding the legitimacy of data use by the AI model (Hayes et al. 2020), the decisions made by the AI model, and the AI model in general (Sætra 2020). As AI models can only base their decisions on the data they were programmed to use, AI models can have a limited perspective (Sadowski et al. 2021). Additionally, the lack of transparency can result in confusion regarding responsibilities when problems arise (Sætra 2020). For example, in a case where a problematic consequence of an AI model emerges within the field of public policy decision-making, it is unclear whether the public agency basing its decisions on the AI model, the service providing the model, or its developers should be held accountable (Paul 2022). In short, AI models and their lack of transparency can obscure biases and abdicate humans from their responsibility to act (Kuziemski and Misuraca 2020), leading to concerns of a lack of democratic control over the development and use of AI (Erman and Furendal 2022).

AI in use

The social and democratic sustainability of AI models in use is frequently questioned in research. On a positive note, authors suggest AI models can play a supportive role for democracy. AI-based technologies can help close the gap between public administration and citizens, using personalization and simplified communication methods, increasing transparency and encouraging participation (Savaget et al. 2019; Anastasiadou et al. 2021; Cavaliere and Romeo 2022). Additionally, AI could increase efficiency in social rights and public policy areas, for example by gaining a better understanding of marginalized groups (Cavaliere and Romeo 2022; Niklas and Dencik 2021).

At the same time, using AI can have socially and democratically unsustainable impacts. First, AI models are not always transparent regarding the flow and use of personal information when it is shared with other actors, which is a privacy infringement (Hayes et al. 2020). As AI requires constant feeding of (new) data, this privacy infringement is continuous. Second, the increased implementation of AI can diminish human autonomy and decision-making power (Arogyaswamy 2020), or influence human actions in such a way that people lose control over their choices and behaviour without sufficient transparency regarding how their behaviour is influenced (Hayes et al. 2020; Milano et al. 2020).

Third, AI models can lead to unequal treatment of citizens (Rafanelli 2022). Partly, this is the result of biases embedded during the AI development, which went unnoticed or was not acted upon. There is, however, another inequality with the use of AI. Often, only a select group of actors, such as governments or larger private companies, are able to reap the benefits of the use of AI, while society as a whole is negatively impacted, for example, by the (mis)use of data or (sector-specific) unemployment (Arogyaswamy 2020; Niklas and Dencik 2021). This enhances the power imbalance between these actors and regular citizens (Erman and Furendal 2022). Governments can, for example, use biometric surveillance systems based on AI models, which threatens the central principles of justice and democracy (Barkane 2022; Ferretti 2022; Kuziemski and Misuraca 2020). Additionally, with the increased use of algorithmic decision-making in public policy and public services, there is also an upsurge of surveillance and scrutiny within the provision of social services, which can be especially damaging for those groups which often need such services (Paul 2022; Gabriel 2022; Niklas and Dencik 2021).

Fourth, AI use has a consequence for access to information. Many digital platforms apply AI to filter information on a personal basis. This means that individuals only receive information in line with their own biases, thus reinforcing these biases with no diverging information (Milano et al. 2020; Savaget et al. 2019). Additionally, citizens are increasingly confronted with disinformation, which damages citizens' trust in the media, problematising the public debate (Vaccari and Chadwick 2020). Furthermore, information platforms use AI to moderate or remove content. This means that AI models used by private companies make controversial moral choices about what constitutes "acceptable" or "unacceptable" speech and, by doing so, set the limits for freedom of speech and pluralism (Ferretti 2022; Marsden et al. 2020; Elkin-Koren 2020). Finally, AI applications are increasingly used in education, the root of the information system. Often used as a way to personalize education, this challenges the values and practices of public and critical democratic education (Saltman 2020).

Supporting infrastructure

Research into the social and democratic impact of AI-supporting infrastructure in the publications included in our dataset is limited. In general, it can be stated that this infrastructure results in unsafe employment, for example in raw material mining and e-waste recycling (Crawford 2021). Additionally, a new inequality presents itself in the raw materials theme: (personal) data has become a 'new' raw material, for which people are seldom compensated. Especially the data of marginalized groups is collected to a greater extent, for example as they make use of social services (Gabriel 2022; Niklas and Dencik 2021).

Coping strategies

As the different governing bodies of many countries grapple with the impact of AI, both ecologically and socially, the governance of AI has become a critical policy area (Kuziemski and Misuraca 2020; Milano et al. 2020). In addition to the general plea for more awareness of the issues mentioned, various publications in our two reviews offer more concrete solutions, which we show in Table 5.1.

In general, calculating the impact of AI is perceived as a first step to reducing this impact, as it can help to understand what strategies are effective. As of yet, there are no standard measurements for the impact of the AI ecosystem on the environment or society (Bloomfield et al. 2021; Gutiérrez et al. 2022; Henderson et al. 2022). Several researchers have attempted to calculate the CO^2 emissions of the computational process of a single AI model, using different proxies for energy (Bloomfield et al. 2021; Dacrema et al. 2019; García-Martín et al. 2019; Henderson et al. 2022; Lacoste et al. 2019; Lannelongue et al. 2021; Schwartz et al. 2020; Strubell et al. 2020). Most investigations however, ended in estimations or prognoses, found it very difficult to estimate future impact or to draw clear boundaries of the impacts to include, and excluded the growing impact of the supporting infrastructure (Hershcovich et al. 2022; Jääskeläinen et al. 2022; Patterson et al. 2021; Probst 2022). Improvement of the measurements is not expected anytime soon (Kar et al. 2022; Robbins and van Wynsberghe 2022; Wu et al. 2022), but different coping strategies can already be discerned from the literature.

AI lifecycle

Coping strategies to mitigate negative impacts of the AI lifecycle cannot be split into AI development and AI in use, as they are strongly interconnected. During AI development, there are several ways to reduce negative ecological impact. Limiting the training time of the model or recycling older models, for example, can significantly reduce the carbon footprint of the model, without necessarily compromising its accuracy (Badar et al. 2021; Sharma et al. 2022). Computationally 'simpler' models might even outperform more complex algorithms (Dacrema et al. 2019; Patterson et al. 2021). There are also strategies to make AI models more energy efficient, even after their initial development (Wu et al. 2022). At the same time, AI models which are less pre-trained, and focus more on learning while in operation, might use less energy during the development phase (Patterson et al. 2021) but require more computational power when in use (Thompson et al. 2022; Wu et al. 2022). The model might require more adjustments, and users may need multiple uses before reaching the desired results (Stoll et al. 2022).

As for the social and democratic sustainability impact, designing-against-discrimination and ethically conscious design could prevent biases, but would mean that developers need to be able to recognize biases when developing AI models (Mantini 2022; Martínez-Ramil 2021). Biases can also be removed after the initial design (Paris 2021; Peters 2022). The removing of bias or adjusting the algorithms for fairness has a caveat: humans often do not recognize 'wrong' decisions by AI models, or are convinced the algorithm is always correct. As AI models lack the contextual awareness to detect, evaluate and make contextually correct decisions, AI requires human intervention, for example to interpret flagged content or to make judiciary decisions (Marsden et al. 2020; Wachter et al. 2021).

To support this, some authors suggest governments should create legal and policy guidance for AI, increasing the requirements for technology corporations which develop AI to act against biases and other harms by their AI technology (Paris 2021; Wachter et al. 2021). This can take the form of human rights impact assessments, disclosure schemes, 'fluid observation' by a critical civil society for responsible innovation, or establishing institutions for public oversight and monitoring of AI (Niklas and Dencik 2021; Buhmann and Fieseler 2021; Smuha 2021). Similar measures could also be used to increase ecological transparency of AI models (Hershcovich et al. 2022). This could also help to clarify responsibilities and accountability in the complex AI landscape.

Furthermore, the current major imbalance in power between citizens and technology corporations and (in particular, authoritarian) governments raises doubts about the ability of citizens to protest implemented AI (Arogyaswamy 2020). Therefore, it remains crucial to be cautious when implementing AI in society (Zimmermann and Lee-Stronach 2022).

Supporting infrastructure

Research regarding coping strategies for the impact of the supporting infrastructure is lacking. The problems are acknowledged, including societal and health problems connected to mining of raw materials and (illegal) dumping of e-waste in (often) low-and-middle-income countries (Samuel and Lucassen 2022), but no clear solutions are put forward to, for example, increase the amount of recycling and the safety of this recycling process (Crawford 2021). It needs to be considered that, regardless of recycling or efficiency, a certain amount of raw materials will always be needed for the AI ecosystem (Cioca and Schuszter 2022; Robbins and van Wynsberghe 2022).

Some coping strategies are proposed in the literature. First, using datacentres instead of on-site hardware can be deemed more resource efficient and can thus be part of a sustainability strategy (Patterson et al. 2021; Wu et al. 2022). Second, datacentres have developed Power Usage Effectiveness (PUE) as a standard metric for measuring the energy usage, specifically computer equipment and ancillary services (e.g., cooling, lighting) (Yang et al. 2022; Masanet et al. 2020). However, although PUE gives some insight, it does not include the energy usage of communication between users and machines, or the energy use of the storage of data (Banet et al. 2021), thus leading to an incomplete picture of the total energy use. Third, the datacentre provider can opt for carbon offset schemes, but the effectiveness of these are debatable (Hershcovich et al. 2022; Weidinger et al. 2021). Finally, although 'green' digital technology strategies are proposed to decrease the need for materials and e-waste, these strategies focus on increasing resource efficiency using hardware and software; very few publications challenge the overconsumption and overproduction of data and the increasing reliance on digital technologies (Samuel and Lucassen 2022).

Discussion

Artificial intelligence is booming. Despite the narrative of hope around the use of AI for sustainability, our literature review has shown that AI also has numerous unsustainable aspects. Apart from the opportunity-synergies, the European Commission aims to harness by coupling the green and digital transition in the twin transition, trade-offs and risk-synergies can also emerge. Therefore, we focus on the following: (1) the interaction within the AI ecosystem, (2) the interaction between the ecological sustainability impact and the social and democratic sustainability impact, and (3) the cross-border effects.

To start, there is the interaction within the AI ecosystem. Coping strategies should take this interaction into account: exclusive focus on reducing the CO_2 emissions of AI development might lead to an increase in CO_2 emissions of AI in use and/or more hardware being deemed obsolete for the sake of maximum efficiency. Therefore, it is important to consider the whole AI ecosystem when setting sustainability standards.

Second, there are interactions between AI's effect on ecological sustainability and social and democratic sustainability. Using the previous example, to reduce the CO_2 emissions of AI development, AI developers could opt to train their AI on smaller datasets, as this consumes less energy and processing power. However, a smaller dataset also often lacks diversity. In use, this AI model is therefore at risk of being biased, thus having a negative social and democratic impact. There could also be a positive synergy: less complex AI models are both more energy efficient and easier to exert democratic control over. It is important to focus on such synergies and attempt to mitigate trade-offs.

Third, the research in our dataset lacks consideration of cross-border effects of the AI ecosystem, yet the impact of AI, and especially its supporting (hardware) infrastructure, on the global environmental and justice issues needs to be reflected upon. Using the previous example, phasing out hardware for maximum software efficiency increases the need for raw materials, and adds to the e-waste problem. In publications concerning the AI ecosystem, research is often focussed on the global north countries, where these problems are less visible (Nost and Colven

2022; Probst 2022). Yet, the sourcing and shipping of rare metals and the conflicts this brings to already fragile states (Church and Crawford 2020; Prause 2020; Robbins and van Wynsberghe 2022) requires a wider re-evaluation of the sustainability of AI and the broader digital infrastructure.

References

Anastasiadou, M., Santos, V., & Montargil, F. (2021). Which technology to which challenge in democratic governance? An approach using design science research. *Transforming Government: People, Process and Policy*, 15(4), 512–531. Scopus. doi: 10.1108/TG-03-2020-0045

Arogyaswamy, B. (2020). Big tech and societal sustainability: An ethical framework. *AI and Society*, 35(4), 829–840. Scopus. doi: 10.1007/s00146-020-00956-6

Asha, P., Mannepalli, K., Khilar, R., Subbulakshmi, N., Dhanalakshmi, R., Tripathi, V., … Sudhakar, M. (2022). Role of machine learning in attaining environmental sustainability. *Energy Reports*, 8, 863–871. doi: 10.1016/j.egyr.2022.09.206

Badar, A., Varma, A., Staniec, A., Gamal, M., Magdy, O., Iqbal, H., … Zonooz, B. (2021, June 6). Highlighting the Importance of Reducing Research Bias and Carbon Emissions in CNNs. *arXiv*. doi: 10.48550/arXiv.2106.03242

Banet, C., Pollitt, M., Covatariu, A., & Duma, D. (2021). *Data Centres and the Grid – Greening ICT in Europe*. Brussels: Centre on Regulation in Europe. Available at Centre on Regulation in Europe website: https://cerre.eu/publications/data-centres-and-the-energy-grid/ (Accessed 11 July 2022).

Bannister, F., & Connolly, R. (2020). Administration by algorithm: A risk management framework. *Information Polity*, 25(4), 471–490. Scopus. doi: 10.3233/IP-200249

Barkane, I. (2022). Questioning the EU proposal for an Artificial Intelligence Act: The need for prohibitions and a stricter approach to biometric surveillance. *Information Polity*, 27(2), 147–162. Scopus. doi: 10.3233/IP-211524

Bender, E. M., Gebru, T., McMillan-Major, A., & Shmitchell, S. (2021). On the Dangers of Stochastic Parrots: Can Language Models Be Too Big? *Proceedings of the 2021 ACM Conference on Fairness, Accountability, and Transparency*, 610–623. New York, NY, USA: Association for Computing Machinery. doi: 10.1145/3442188.3445922

Bloomfield, P., Clutton-Brock, P., Pencheon, E., Magnusson, J., & Karpathakis, K. (2021). Artificial intelligence in the NHS: Climate and emissions. *The Journal of Climate Change and Health*, 4, 100056. doi: 10.1016/j.joclim.2021.100056

Borowiec, D., Yeung, G., Friday, A., Harper, R. H. R., & Garraghan, P. (2022). Trimmer: Cost-Efficient Deep Learning Auto-tuning for Cloud Datacenters. *IEEE 15th International Conference on Cloud Computing (CLOUD)*, 374–384. doi: 10.1109/CLOUD55607.2022.00061

Brevini, B. (2020). Black boxes, not green: Mythologizing artificial intelligence and omitting the environment. *Big Data & Society*, 7(2), 2053951720935141. doi: 10.1177/2053951720935141

Bucknall, B. S., & Dori-Hacohen, S. (2022). Current and Near-Term AI as a Potential Existential Risk Factor. *Proceedings of the 2022 AAAI/ACM Conference on AI, Ethics, and Society*, 119–129. New York, NY, USA: Association for Computing Machinery. doi: 10.1145/3514094.3534146

Budennyy, S. A., Lazarev, V. D., Zakharenko, N. N., Korovin, A. N., Plosskaya, O. A., Dimitrov, D. V., … Zhukov, L. E. (2022). eco2AI: Carbon emissions tracking of machine learning models as the first step towards sustainable AI. *Doklady Mathematics*, 106(1), S118–S128. doi: 10.1134/S1064562422060230

Buhmann, A., & Fieseler, C. (2021). Towards a deliberative framework for responsible innovation in artificial intelligence. *Technology in Society*, 64. Scopus. doi: 10.1016/j.techsoc.2020.101475

Cavaliere, P., & Romeo, G. (2022). From poisons to antidotes: Algorithms as democracy boosters. *European Journal of Risk Regulation*, 13(3), 421–442. Scopus. doi: 10.1017/err.2021.57

Cherubini, E., Zanghelini, G. M., Piemonte, D., Muller, N. B., Dias, R., Kabe, Y. H. O., & Soto, J. (2019). Environmental sustainability for highways operation: Comparative analysis of plastic and steel screen anti-glare systems. *Journal of Cleaner Production*, 240, 118152. doi: 10.1016/j.jclepro.2019.118152

Church, C., & Crawford, A. (2020). Minerals and the metals for the energy transition: Exploring the conflict implications for mineral-rich, fragile states. In M. Hafner & S. Tagliapietra (Eds.), *The Geopolitics of the Global Energy Transition* (pp. 279–304). Cham: Springer International Publishing.

Cioca, M., & Schuszter, I. C. (2022). A system for sustainable usage of computing resources leveraging deep learning predictions. *Applied Sciences*, 12(17), 8411. https://doi.org/10.3390/app12178411

Clutton-Brock, P., Rolnick, D., Donti, P. L., & Kaack, L. (2021). *Climate Change and AI. Recommendations for Government Action*. GPAI, Climate Change AI, Centre for AI & Climate.

Crawford, K. (2021). *Atlas of AI: The Real Worlds of Artificial Intelligence: Power, Politics, and the Planetary Costs of Artificial Intelligence*. New Haven, CT: Yale University Press.

Dacrema, M. F., Cremonesi, P., & Jannach, D. (2019). Are we really making much progress? A worrying analysis of recent neural recommendation approaches. *Proceedings of the 13th ACM Conference on Recommender Systems*, 101–109. doi: 10.1145/3298689.3347058

Dauvergne, P. (2022). Is artificial intelligence greening global supply chains? Exposing the political economy of environmental costs. *Review of International Political Economy*, 29(3), 696–718. doi: 10.1080/09692290.2020.1814381

Elkin-Koren, N. (2020). Contesting algorithms: Restoring the public interest in content filtering by artificial intelligence. *Big Data and Society*, 7(2). Scopus. https://doi.org/10.1177/2053951720932296

Elzen, B., & Wieczorek, A. (2005). Transitions towards sustainability through system innovation. *Technological Forecasting and Social Change*, 72(6), 651–661. doi: 10.1016/j.techfore.2005.04.002

Erman, E., & Furendal, M. (2022). The global governance of artificial intelligence: Some normative concerns. *Moral Philosophy and Politics*, 9(2), 267–291. Scopus. https://doi.org/10.1515/mopp-2020-0046

European Commission. (2021). 'Fit for 55': Delivering the EU's 2030 Climate Target on the way to climate neutrality. *European Commission*. Available at https://ec.europa.eu/info/strategy/priorities-2019-2024/european-green-deal/delivering-european-green-deal_en (Accessed 18 July 2021).

European Commission, Joint Research Centre, Muench, S., Stoermer, E., Jensen, K., Asikainen, T., … Scapolo, F. (2022a). Towards a green & digital future: Key requirements for successful twin transitions in the European Union. *Luxembourg: Joint Research Centre European Commission. Publications Office of the European Union*. Available at https://data.europa.eu/doi/10.2760/977331 (Accessed 25 July 2022).

European Commission. (2022b). Communication from the Commission to the European Parliament and the Council 2022 Strategic Foresight Report Twinning the green and digital transitions in the new geopolitical context. Pub. L. No. COM/2022/289 final, 2022.

European Commission. (2022c). The European Commission's priorities [Text]. Available at European Commission website: https://ec.europa.eu/info/strategy/priorities-2019-2024_en (Accessed 1 August 2022).

Ferretti, T. (2022). An institutionalist approach to AI Ethics: Justifying the priority of government regulation over self-regulation. *Moral Philosophy and Politics*, 9(2), 239–265. Scopus. https://doi.org/10.1515/mopp-2020-0056

Gabriel, I. (2022). Toward a theory of justice for artificial intelligence. *Daedalus*, 151(2), 218–231. Scopus. https://doi.org/10.1162/daed_a_01911

García-Martín, E., Rodrigues, C. F., Riley, G., & Grahn, H. (2019). Estimation of energy consumption in machine learning. *Journal of Parallel and Distributed Computing*, 134, 75–88. doi: 10.1016/j.jpdc.2019.07.007

Gomez, A., Tretter, A., Hager, P. A., Sanmugarajah, P., Benini, L., & Thiele, L. (2022). Dataflow driven partitioning of machine learning applications for optimal energy use in batteryless systems. *ACM Transactions on Embedded Computing Systems*, 21(5), 54:1–54:29. https://doi.org/10.1145/3520135

Greenhouse Gas Protocol. (2013). *Technical Guidance for Calculating Scope 3 Emissions*. Geneva: World Resources Institute & World Business Council for Sustainable Development.

Gutiérrez, M., Moraga, M. Á., & García, F. (2022). Analysing the energy impact of different optimisations for machine learning models. *2022 International Conference on ICT for Sustainability (ICT4S)*, 46–52. doi: 10.1109/ICT4S55073.2022.00016

Halsband, A. (2022). Sustainable AI and Intergenerational Justice. *Sustainability (Switzerland)*, 14(7). Scopus. https://doi.org/10.3390/su14073922

Hayes, P., van de Poel, I., & Steen, M. (2020). Algorithms and values in justice and security. *AI and Society*, 35(3), 533–555. Scopus. https://doi.org/10.1007/s00146-019-00932-9

Henderson, P., Hu, J., Romoff, J., Brunskill, E., Jurafsky, D., & Pineau, J. (2022, November 29). Towards the systematic reporting of the energy and carbon footprints of machine learning. *arXiv*. doi: 10.48550/arXiv.2002.05651

Hershcovich, D., Webersinke, N., Kraus, M., Bingler, J. A., & Leippold, M. (2022, October 18). Towards climate awareness in NLP research. *arXiv*. doi: 10.48550/arXiv.2205.05071

Jääskeläinen, P., Pargman, D., & Holzapfel, A. (2022, October 3). Towards sustainability assessment of artificial intelligence in artistic practices. *arXiv*. doi: 10.48550/arXiv.2210.08981

Kaack, L. H., Donti, P. L., Strubell, E., Kamiya, G., Creutzig, F., & Rolnick, D. (2022). Aligning artificial intelligence with climate change mitigation. *Nature Climate Change*, 12(6), 518–527. doi: 10.1038/s41558-022-01377-7

Kalogirou, S. A. (2007). Introduction to artificial intelligence technology. In S. A. Kalogirou (Ed.), *Artificial Intelligence in Energy and Renewable Energy Systems* (pp. 1–46). New York: Nova Science Publishers.

Kar, A. K., Choudhary, S. K., & Singh, V. K. (2022). How can artificial intelligence impact sustainability: A systematic literature review. *Journal of Cleaner Production*, 376, 134120. doi: 10.1016/j.jclepro.2022.134120

Kopka, A., & Grashof, N. (2022). Artificial intelligence: Catalyst or barrier on the path to sustainability? *Technological Forecasting and Social Change*, 175, 121318. doi: 10.1016/j.techfore.2021.121318

Kum, S., Oh, S., Yeom, J., & Moon, J. (2022). Optimization of edge resources for deep learning application with batch and model management. *Sensors*, 22(17), 6717. doi: 10.3390/s22176717

Kuziemski, M., & Misuraca, G. (2020). AI governance in the public sector: Three tales from the frontiers of automated decision-making in democratic settings. *Telecommunications Policy*, 44(6). Scopus. doi: 10.1016/j.telpol.2020.101976

Lacoste, A., Luccioni, A., Schmidt, V., & Dandres, T. (2019, November 4). Quantifying the carbon emissions of machine learning. *arXiv*. doi: 10.48550/arXiv.1910.09700

Lannelongue, L., Grealey, J., & Inouye, M. (2021). Green algorithms: Quantifying the carbon footprint of computation. *Advanced Science*, 8(12), 2100707. doi: 10.1002/advs.202100707

Mantini, A. (2022). Technological sustainability and artificial intelligence algor-ethics. *Sustainability (Switzerland)*, 14(6). Scopus. doi: 10.3390/su14063215

Marsden, C., Meyer, T., & Brown, I. (2020). Platform values and democratic elections: How can the law regulate digital disinformation? *Computer Law and Security Review*, 36. Scopus. doi: 10.1016/j.clsr.2019.105373

Martínez-Ramil, P. (2021). Is the EU human rights legal framework able to cope with discriminatory AI? *Revista de Internet, Derecho y Politica*, (34). Scopus. doi: 10.7238/idp.v0i34.387481

Masanet, E., Shehabi, A., Lei, N., Smith, S., & Koomey, J. (2020). Recalibrating global data center energy-use estimates. *Science*, 367(6481), 984–986. doi: 10.1126/science.aba3758

Milano, S., Taddeo, M., & Floridi, L. (2020). Recommender systems and their ethical challenges. *AI and Society*, 35(4), 957–967. Scopus. doi: 10.1007/s00146-020-00950-y

Niklas, J., & Dencik, L. (2021). What rights matter? Examining the place of social rights in the eu's artificial intelligence policy debate. *Internet Policy Review*, 10(3). Scopus. doi: 10.14763/2021.3.1579

Nilashi, M., Rupani, P. F., Rupani, M. M., Kamyab, H., Shao, W., Ahmadi, H., … Aljojo, N. (2019). Measuring sustainability through ecological sustainability and human sustainability: A machine learning approach. *Journal of Cleaner Production*, 240, 118162. doi: 10.1016/j.jclepro.2019.118162

Nordgren, A. (2022). Artificial intelligence and climate change: Ethical issues. *Journal of Information, Communication and Ethics in Society*, 21(1), 1–15. doi: 10.1108/JICES-11-2021-0106

Nost, E., & Colven, E. (2022). Earth for AI: A political ecology of data-driven climate initiatives. *Geoforum*, 130, 23–34. doi: 10.1016/j.geoforum.2022.01.016

Page, M. J., McKenzie, J. E., Bossuyt, P. M., Boutron, I., Hoffmann, T. C., Mulrow, C. D., … Moher, D. (2021). The PRISMA 2020 statement: An updated guideline for reporting systematic reviews. *Systematic Reviews*, 10(1), 89. doi: 10.1186/s13643-021-01626-4

Paris, B. (2021). Configuring fakes: Digitized bodies, the politics of evidence, and agency. *Social Media and Society*, 7(4). Scopus. doi: 10.1177/20563051211062919

Patterson, D., Gonzalez, J., Le, Q., Liang, C., Munguia, L.-M., Rothchild, D., … Dean, J. (2021, April 23). Carbon emissions and large neural network training. *arXiv*. doi: 10.48550/arXiv.2104.10350

Paul, R. (2022). Can critical policy studies outsmart AI? Research agenda on artificial intelligence technologies and public policy. *Critical Policy Studies*, 16(4), 497–509. Scopus. doi: 10.1080/19460171.2022.2123018

Peters, U. (2022). Algorithmic political bias in artificial intelligence systems. *Philosophy and Technology*, 35(2). Scopus. doi: 10.1007/s13347-022-00512-8

Poole, D., & Mackworth, A. (2010). Artificial intelligence and agents. In D. Poole & A. Mackworth (Eds.), *Artificial Intelligence: Foundations of Computational Agents* (pp. 3–42). Cambridge: Cambridge University Press.

Prause, L. (2020). Chapter 10 - Conflicts related to resources: The case of cobalt mining in the democratic republic of Congo. In A. Bleicher & A. Pehlken (Eds.), *The Material Basis of Energy Transitions* (pp. 153–167). Academic Press. doi: 10.1016/B978-0-12-819534-5.00010-6

Probst, D. (2022, October 1). Social and environmental impact of recent developments in machine learning on biology and chemistry research. *arXiv*. doi: 10.48550/arXiv.2210.00356

Rafanelli, L. M. (2022). Justice, injustice, and artificial intelligence: Lessons from political theory and philosophy. *Big Data and Society*, 9(1). Scopus. doi: 10.1177/20539517221080676

Robbins, S., & van Wynsberghe, A. (2022). Our new artificial intelligence infrastructure: Becoming locked into an unsustainable future. *Sustainability (Switzerland)*, 14(8). Scopus. doi: 10.3390/su14084829

Rohde, F., Gossen, M., Wagner, J., & Santarius, T. (2021). Sustainability challenges of artificial intelligence and policy implications. Ökologisches Wirtschaften - Fachzeitschrift, 36(O1), 36–40. doi: 10.14512/OEWO360136

Rosário, A. T., & Dias, J. C. (2022). Sustainability and the digital transition: A literature review. *Sustainability*, 14(7), 4072. doi: 10.3390/su14074072

Sadowski, J., Viljoen, S., & Whittaker, M. (2021). Everyone should decide how their digital data are used—Not just tech companies. *Nature*, 595(7866), 169–171. Scopus. doi: 10.1038/d41586-021-01812-3

Sætra, H. S. (2020). A shallow defence of a technocracy of artificial intelligence: Examining the political harms of algorithmic governance in the domain of government. *Technology in Society*, 62. Scopus. doi: 10.1016/j.techsoc.2020.101283

Saltman, K. J. (2020). Artificial intelligence and the technological turn of public education privatization: In defence of democratic education. *London Review of Education*, 18(2), 196–208. Scopus. doi: 10.14324/LRE.18.2.04

Samuel, G., & Lucassen, A. M. (2022). The environmental sustainability of data-driven health research: A scoping review. *Digital Health*, 8. doi: 10.1177/20552076221111297

Sarangi, S., & Sharma, P. (2019). *Artificial Intelligence: Evolution, Ethics and Public Policy*. London: Routledge.

Savaget, P., Chiarini, T., & Evans, S. (2019). Empowering political participation through artificial intelligence. *Science and Public Policy*, 46(3), 369–380. Scopus. doi: 10.1093/scipol/scy064

Schwartz, R., Dodge, J., Smith, N. A., & Etzioni, O. (2020). Green AI. *Communications of the ACM*, 63(12), 54–63. doi: 10.1145/3381831

Shahid, A., & Mushtaq, M. (2020). A survey comparing specialized hardware and evolution In TPUs for neural networks. *IEEE 23rd International Multitopic Conference (INMIC)*, 1–6. doi: 10.1109/INMIC50486.2020.9318136

Sharma, M., Mishra, T. K., & Kumar, A. (2022). Source code auto-completion using various deep learning models under limited computing resources. *Complex & Intelligent Systems*, 8(5), 4357–4368. doi: 10.1007/s40747-022-00708-7

Smuha, N. A. (2021). Beyond the individual: Governing ai's societal harm. *Internet Policy Review*, 10(3). Scopus. doi: 10.14763/2021.3.1574

Sovacool, B. K., Burke, M., Baker, L., Kotikalapudi, C. K., & Wlokas, H. (2017). New frontiers and conceptual frameworks for energy justice. *Energy Policy*, 105, 677–691.

Starke, C., Baleis, J., Keller, B., & Marcinkowski, F. (2022). Fairness perceptions of algorithmic decision-making: A systematic review of the empirical literature. *Big Data and Society*, 9(2). Scopus. doi: 10.1177/20539517221115189

Stoll, C., Gallersdörfer, U., & Klaaßen, L. (2022). Climate impacts of the metaverse. *Joule*, 6(12), 2668–2673. doi: 10.1016/j.joule.2022.10.013

Stray, J., Vendrov, I., Nixon, J., Adler, S., & Hadfield-Menell, D. (2021, July 22). What are you optimizing for? Aligning recommender systems with human values. *arXiv*. doi: 10.48550/arXiv.2107.10939

Strubell, E., Ganesh, A., & McCallum, A. (2020). Energy and policy considerations for modern deep learning research. *Proceedings of the AAAI Conference on Artificial Intelligence*, 34(09), 13693–13696. doi: 10.1609/aaai.v34i09.7123

Thompson, N. C., Greenewald, K., Lee, K., & Manso, G. F. (2022, July 27). The computational limits of deep learning. *arXiv*. doi: 10.48550/arXiv.2007.05558

Vaccari, C., & Chadwick, A. (2020). Deepfakes and disinformation: Exploring the impact of synthetic political video on deception, uncertainty, and trust in news. *Social Media and Society*, 6(1). Scopus. doi: 10.1177/2056305120903408

Wachter, S., Mittelstadt, B., & Russell, C. (2021). Why fairness cannot be automated: Bridging the gap between EU non-discrimination law and AI. *Computer Law and Security Review*, 41. Scopus. doi: 10.1016/j.clsr.2021.105567

Weidinger, L., Mellor, J., Rauh, M., Griffin, C., Uesato, J., Huang, P.-S., … Gabriel, I. (2021, December 8). Ethical and social risks of harm from Language Models. *arXiv*. doi: 10.48550/arXiv.2112.04359

Wu, C.-J., Raghavendra, R., Gupta, U., Acun, B., Ardalani, N., Maeng, K., … Hazelwood, K. (2022). Sustainable AI: Environmental implications, challenges and opportunities. *Proceedings of Machine Learning and Systems*, 4, 795–813.

Yang, Z., Du, J., Lin, Y., Du, Z., Xia, L., Zhao, Q., & Guan, X. (2022). Increasing the energy efficiency of a data center based on machine learning. *Journal of Industrial Ecology*, 26(1), 323–335. doi: 10.1111/jiec.13155

Završnik, A. (2020). Criminal justice, artificial intelligence systems, and human rights. *ERA Forum*, 20(4), 567–583. Scopus. doi: 10.1007/s12027-020-00602-0

Zimmermann, A., & Lee-Stronach, C. (2022). Proceed with caution. *Canadian Journal of Philosophy*, 52(1), 6–25. Scopus. doi: 10.1017/can.2021.17

6

PROBLEMATISING DIGITAL DEMOCRACY

The role of context in shaping digital participation

Caitlin Hafferty, Jiří Pánek, and Ian Babelon

Introduction

Digital technology continuously transforms participatory processes to achieve just and sustainable futures in research, policy, and practice (Afzalan and Muller, 2018; Evans-Cowley and Hollander, 2010; Hafferty, 2022; Wilson and Tewdwr-Jones, 2021). While international strategies for digital transformation emphasise the role of technology in enhancing democratic participation, there are increasing concerns about participation gaps such as digital literacy and access to the internet as barriers to equitable and inclusive decision-making. Digital inequalities and exclusions risk jeopardising just sustainability which seeks to embed social justice at the heart of sustainability transitions. Despite touted benefits, questions persist about the effectiveness of digital technology for addressing the goals of participation and the extent of negative impacts, particularly the exacerbation and emergence of new social justice issues and democratic dilemmas. This chapter explores the contextual factors and socio-economic dynamics that shape digital participation's benefits and pitfalls for achieving just and sustainable outcomes. Drawing from a critical interpretive review (e.g., Hirons, 2021), we synthesise recent research on digital participatory planning and environmental decision-making in the UK and globally. In doing so, we explore how the contextual factors and socio-economic dynamics that influence outcomes in all participatory processes (regardless of the digital, in-person, or hybrid methods used) take on unique features in digital and remote environments. Insights highlight a variety of key considerations regarding the effectiveness of technology for fostering meaningful and effective participation, encompassing the impact of digital tools on inclusion, social and contextual cues, power, and trust. Challenging attitudes of "digital-by-default" and a "blind-faith optimism" in technology (Charlton et al., 2023), we argue for more adaptable

DOI: 10.4324/9781003441311-7

hybrid approaches over a one-size-fits-all approach. This chapter underscores that there are no purely technological solutions to complex sustainability and social justice challenges.

Participation in an increasingly digitised world

In this chapter, *participation* is considered to include ways of effectively and ethically engaging people in the processes, structures, spaces, and decisions that affect their lives, actively working with them to achieve equitable, sustainable, and socially just outcomes (Kindon, 2009). *Digital* is considered a format or process, including technology that uses, stores, and processes data or information in the form of digital signals, which can be either online or offline, synchronous (taking place in real time) or asynchronous (conducted remotely and at different times). This chapter focuses on *digital and remote* participation. Digital participation has been increasingly institutionalised in policy and practice, frequently accompanied by optimistic narratives about the opportunities that digital transitions provide for supporting more democratic, transparent, and inclusive decision-making across governments, private industry, academic institutions, and a range of other stakeholders (Charlton et al., 2023; Hafferty, 2022).

National and international strategies for digital transformation often frame digitalisation as a "win-win" for both people and planet, emphasising the possibilities of technological innovation for enabling more sustainable and equitable futures. For example, the European Union's Digital Strategy (European Commission, 2023a) and Digital Democracy Initiative (European Commission, 2023b) aim to tap into the vast opportunities offered by Web 4.0 (characterised by the integration of digital and real objects and environments, with enhanced interactions between humans and machines) and virtual worlds to shape a more human-centred, sustainable future which empowers citizens while promoting democracy and human rights with digital technologies. The UN integrates (digital) participation into two Sustainable Development Goals (SDGs) (see United Nations, 2023). SDG 11 focuses on creating inclusive, safe, resilient, and sustainable cities. It includes Target 11.3, aiming for inclusive and sustainable urbanisation through Indicator 11.3.2, which promotes participatory urban planning by 2030. SDG 16 promotes peaceful societies with accessible justice and inclusive institutions. Target 16.7 commits to responsive, inclusive, and participatory decision-making at all levels. Relevant to participation discussions, Indicator 16.7.2 gauges public perception of inclusive decision-making across diverse groups (e.g., age, disability, and socio-economic status).

In the UK, strategies for simultaneous digital transformation and economic growth have followed a 'digital-by-default' or 'digital first' mantra in line with the Government's aspirations to become a world leader in digital adoption (Cabinet Office, 2012; DDCMS, 2022; DLUHC, 2022; Government Digital Service, 2017). Digital-by-default remains a strong narrative across the UK public sector from national to local levels, for example, the Digital Leader's Public Sector Innovation Conference (2022) included a strong focus on the efficiency, interoperability, and 'limitless potential' of technology and data for building digitally enabled public services, as well as the 'accelerated digital transformation' and 'rapid move towards digital first'. Across the globe, an abundance of online platforms for public and stakeholder participation have been identified – including *Bang the Table – Engagement HQ, Maptionnaire, Commonplace, and CitizenLab* – which claim to promote more accessible and inclusive citizen engagement and public services, foster efficient collaboration between multiple stakeholders, and ensure more transparent and trustworthy decision-making processes (Babelon, 2021; Babelon et al., 2016; Falco and Kleinhans, 2018).

However, in the current policy-driven push towards digital transformation in the UK and internationally, critics warn that "blind faith" and optimism in digital technologies risk undermining democratic processes and decisions (Bernholz et al., 2021; Charlton et al., 2023). Similar warnings were also raised regarding the over-enthusiastic use of geographic information systems in the 1990s (Pickles, 1995). Despite considerable optimism regarding the potential of digital technologies, one key issue is the lack of attention paid to the wider societal implications of rapid and unregulated digital transformation. While national strategies often overlook some important ethical issues regarding digital and data-driven technologies, in contrast, the research indicates that there are significant ethical risks for society including bias and the exacerbation of existing exclusions, injustices, and prejudices (e.g., O'Neil, 2016; Tsamados et al., 2021). Many guidelines, toolkits, and frameworks for engagement also do not sufficiently consider the technical and ethical debates around digital tools and technologies in this context (e.g., recent Organisation for Economic Co-operation and Development (OECD) guidelines for participation includes a limited and insufficient discussion of the challenges and opportunities for digital engagement; see OECD, 2022).

In an increasingly digitised world, there are many unresolved questions about the benefits of these technologies and their effectiveness at addressing the goals of participation (Afzalan and Muller, 2018; Köpsel et al., 2021; Manderscheid et al., 2022; McKinley et al., 2021; Robinson and Johnson, 2021; Willis et al., 2021). Although digital tools have the potential to enhance participation on the one hand, technology, on the other hand, can cause an array of negative consequences which can lead to the (further) exclusion, marginalisation, and disempowerment of participants. In an increasingly digitised world, it is critical that engagement and other participatory processes remain inclusive (particularly of marginalised groups and individuals), meaningful, and achieve intended positive outcomes. While national and international strategies emphasise the importance of digital technologies for enabling effective democratic participation and societal transformations, numerous participation gaps have been highlighted which present barriers to inclusive and equitable decision-making and pose a threat to just sustainability (Agyeman, 2008; Agyeman et al., 2003; Bennett et al., 2019). The digital world is also marked with considerable power imbalances due to the domination of large technology companies (e.g., Amoore, 2020; Bartlett, 2018; Bernholz et al., 2021; Brossard, 2019; McIlwain, 2020; O'Neil, 2016; Wachter et al., 2020) and it is important to remember that the challenges and opportunities for digital engagement are operating against a backdrop of continuous struggles for a more inclusive, democratic, and sustainable society (Certomà, 2021; McLean, 2019, 2020).

The extent to which digital tools lead to more effective, inclusive and meaningful outcomes remains unclear. In order to expose the (un)sustainabilities and injustices of digital participation, this chapter aims to explore the contextual and socio-economic factors which promote (or inhibit) more sustainable and equitable outcomes in environmental decision-making processes. To explore these dynamics in depth, we draw from a critical interpretive review (Hirons, 2021, Rowe, 2014) which aims to synthesise recent work on digital participatory and environmental decision-making in the UK and internationally. Our review spanned a broad range of relevant contexts including natural resource management, conservation, urban and environmental planning, and nature recovery, focusing on the technical barriers and ethical debates around the increasing digitalisation of the ways in which people are included (or excluded) in these spaces.

Digital participatory tools and platforms

There is an abundance of digital tools for participatory planning and environmental decision-making. These tools are used for a wide variety of purposes and at different stages in decision-making processes, from problem exploration and scoping, to feedback on proposals and project evaluation (Møller and Olafsson, 2018). Table 6.1 provides an illustrative, rather than comprehensive, list of digital tools for engagement alongside examples of their use. Digital tools and platforms are used for a wide range of purposes and in a variety of contexts in planning and development, healthcare, and technological innovation (e.g., see Babelon, 2021; Babelon et al., 2021; Hafferty, 2022; Hafferty et al., 2024; Pánek, 2016; Rawat and Yusuf, 2020; Wilson and Tewdwr-Jones, 2021). Digital participatory technologies are often categorised from 'leading' to 'enabling' participation (Arnstein, 1969), for example, social media and communications technologies (Table 6.1) may be more suitable to information sharing rather than two-way deliberation and collaboration (also see Hafferty et al., 2024). To further illustrate how a range of digital tools for participation can be used along decision-making and project lifecycles, with different challenges and opportunities, the remainder of this section provides more detailed descriptions of digital participation in two different contexts: France and the Czech Republic. Both of these examples focus on participation in planning and placemaking, which involve implementing decision-making systems which aim to tackle interconnected social, economic, and environmental challenges through community services, land use decisions, natural resource management,

TABLE 6.1 Digital participatory tools and technologies: examples in planning and environmental decision-making

Digital tools for participation	*Example*
Digital participatory platforms (DPPs)	Babelon (2021), Falco and Kleinhans (2018).
Participatory mapping (e.g., public participation GIS and volunteered GIS)	Brown et al. (2020), Hacklay (2013), Panek (2016), Rawat and Yusuf (2019).
Geo-visualisation (e.g., planning support systems, digital twins, and 3D visualisation)	Kahila-Tani et al. (2016)
Social media, communications software (e.g., webinars and videos), mobile applications	Afzalan and Evans-Cowley (2015); Evans-Cowley (2010).
Collaboration tools (e.g., decision support systems and networking tools)	Carver et al. (2001); Fox et al. (2022).
Gamification (e.g., virtual and augmented reality, serious games, and Minecraft)	Delaney (2022); Galeote et al. (2021).
Open data, information and e-government (e.g., government websites, open databases and dashboards, and interactive maps)	Rawat and Yusuf (2019); Conroy and Evans-Cowley (2006).

the development of green infrastructure, and promotion of more sustainable and resilient local economies.

In the realms of town planning and placemaking, hybrid technology and methods have been instrumental to actively engaging citizens and other stakeholders in the life cycle of planning policies and projects, while helping to transform ways of working among staff at sponsoring organisations. For example, the map-based survey tool Carticipe-Debatomap was used to engage residents across Grenoble Metropolitan Agency (France) for its local development plan update between 2016 and 2018, using an iterative process of both online engagement and via in-person design feedback workshops (Babelon, 2021). Consultants and trained volunteers performed active outreach in the streets to promote and capture live participation on the Carticipe portal, as well as offering experimental game-based approaches to participation. The workshops both discussed the input on the portal and encouraged further online participation after the in-person events. The tool was used in two strategic stages: initially to identify priorities across the metropolitan region, followed by more specific feedback about policy orientations that were largely shaped by the first phase of the consultation. The overall approach, while modest in absolute scale, was unprecedented as a capacity-building effort for the relatively new metropolitan agency, and helped to promote and explain a metropolitan local development plan to residents, which is an otherwise highly complex planning instrument that people typically find difficult to relate to. This two-stage approach to digital public consultation has proven popular and effective across different regions and contexts, from consultations about strategic planning level to the design of public realm improvements and active mobility infrastructure (Babelon, 2021). Key enablers include the readiness to engage both 'deeply' (i.e. through active, meaningful participation in decision-making processes) and 'broadly' (i.e. actively reaching out to diverse groups and publics), as implemented through clear and workable means (see Charlton et al., 2023; Nabatchi and Leighninger, 2015). Accordingly, some city councils have developed their very community engagement strategy *through* community engagement, leading to a clear statement of commitments and limitations, based on an appropriate use of both digital and in-person/physical methods. Among many others, examples include the cities of Longmont and Boulder (Colorado, United States), as well as the Grenoble City Council (France). In spite of promising continuous innovation in the field, participation in planning – regardless of the methods used – always runs the risk of being applied as a consultative, box-ticking exercise with no compelling or evident influence on actual decisions (Kahila-Tani et al., 2019).

In the Czech Republic, participatory planning is developed often by third parties, in this case, the Czech Healthy Cities Network. The network supports its cities to achieve the highest status of the Czech Republic's Government Council for Sustainable Development for Local Agenda 21 work. The network's involvement with the programme (LA21) also resulted in the inclusion of a health dimension to the national assessment system. The network co-develops (with Palacky University Olomouc) the participatory mapping platform Emotional Maps (https://www.pocitovemapy.cz/index-en.html) that can be used by its members for various participatory mapping endeavours (also see Pánek, 2018, 2019). Emotional Maps is an online crowdsourcing tool designed as a web application. It is based on the Leaflet library and allows users to collect spatial data on a map treasure trove similar to other crowdsourced web mapping tools. Unlike Ushahidi, Umap, ArcGIS Online, and many others, it does not require registration or installation of any special software, plug-in or virtual server. Although this may give the impression that this approach does not allow user verification and is susceptible to possible attempts to influence the results, the authors are aware of this weakness and work with it as an additional potential for the development of the application. Since 2014, Emotional Maps have been deployed in over 200 municipalities in several countries. The majority (80%) of the maps were created in the Czech Republic, with Slovakia being the second most common country where the app was used (15%). In other countries (Iceland, the Netherlands, Norway, Serbia, Spain, the United States, etc.), there were usually one-off deployments. The number of questions on each map varied according to their focus and ranged from single-question maps to maps with more than 20 questions. On average, each map contained six spatial questions. In the beginning, the approach was mainly analogue-based, but throughout the time, the developers together with the Healthy Cities Network turned more to the digital approach, yet still combined with paper-based maps. Since almost every map was/is unique, with several hundred questions used on the emotional maps;

however, these can be divided into five main categories cross-cutting numerous socio-economic and environmental challenges:

- Neglected environment (22.1% of all questions)
  - Questions: where is the neglected environment? Where would you appreciate more green space? etc.
- Pleasant environment (21.1% of all questions)
  - Questions: Where do you like it? Where are you proud of the city? Where is a pleasant environment for you? etc.
- Safety (18.8% of all questions)
  - Where do you feel unsafe (day, night)? Where would you appreciate more public lighting? etc.
- Transport (16.8% of all questions)
  - Where do you think there are traffic hazards? Where do you have a problem with parking? Where can't you get to using public transport? etc.
- Leisure and sport (16.6% of all questions)
  - Where do you like to spend your free time? Where do you play sports? Where do you relax?

The role of context in shaping outcomes in participatory processes

Context matters for any participatory decision-making process. Despite bold claims about the potential of digital tools, their success (and harms) varies depending on the (usually local) context and socio-economic dynamics of the environment in which they are used (see Baker and Chapin, 2018; Hafferty, 2022; Reed et al., 2018). Typologies and models for participation can help frame the debates around whether engagement achieves its aspired goals or benefits, or succumbs to its pitfalls, by providing a structure to understand how the dynamic interactions between contextual factors and socio-economic dimensions shape the outcomes (e.g., Arnstein, 1969; IAP2, 2018; see https://participedia.net/ for a crowdsourced repository of various models and frameworks for participation). For example, Reed et al.'s (2018) *theory of participation* – which was developed within the context of conservation and environmental management decision-making – explains how different engagement approaches, methods, and tools are "fit for purpose" while considering participatory rationales, socio-economic context, power relations, design factors, variance over space and time scales, and institutional fit, among other factors. The key contribution of this theory (and others which similarly highlight the role of context in shaping outcomes in participatory decision-making, e.g., Baker and Chapin, 2018; Wesselink et al., 2011) is that there are a range of factors that explain much of the variation in the outcomes of participatory processes (i.e. whether participation achieves more positive or negative outcomes for participants, for society, and for the environment). It is important to be clear that these contextual factors and socio-economic dynamics are inherent to debates around *any* participatory process (see Cooke and Kothari, 2001; Kesby et al., 2007; Stirling, 2008). However, the extent to which these factors hold for digital and remote participation compared to in-person counterparts, or whether there are new considerations for participation in the digital age, remains under-explored (Afzalan and Muller, 2018; Charlton et al., 2023; Hafferty, 2022).[1]

New dimensions in a digital world: the ethical considerations and societal implications of digital participation

A rapidly expanding body literature on digital engagement suggests that the contextual factors that shape outcomes in participatory governance processes take on new dimensions within digital and remote contexts (Afzalan and Muller, 2018; Hafferty, 2022; Hafferty et al., 2024; Panchyshyn and Corbett, 2022; Willis et al., 2021). In the digital age, there are still many unanswered questions about the benefits of digital tools and their effectiveness at addressing the goals of engagement (Afzalan and Muller, 2018; Evans-Cowley and Hollander, 2010).

The COVID-19 pandemic accelerated the adoption and implementation of digital technology in the UK and internationally, which has added urgency to the question of whether meaningful and effective engagement can be conducted in remote and digital settings. For example, in the Czech Republic, the two largest providers of mobile applications for direct communication between municipalities and the citizens reported a 383% increase in 2020 compared with 2019 in the number of municipalities using their services (Panek et al., 2022). While digital technology can enhance engagement on one hand, it can (further) exclude, marginalise, and disempower individuals and groups of people on the other. The remainder of this section explores how the factors that shape outcomes in digital participatory processes can take on new dimensions in entirely digital and remote environments, focusing on a range of technical and ethical issues that may arise. This includes how power imbalances, exclusions, and inequalities manifest in different ways in digital (compared to in-person) participatory processes. To provide structure for this review, we focus on the impact of digital tools on four interlinked factors which have been shown to underpin and shape outcomes in participatory processes – inclusion, social and contextual cues, power, and trust (Baker and Chapin, 2018; Reed et al., 2018; Rowe and Gammack, 2004; Stirling, 2008; Wesselink et al., 2011). The literature suggests that although these factors are fundamental issues for *any* participatory process (regardless of the method used), there is evidence to suggest that there are unique considerations for digital participation, which must be recognised and explicitly addressed in an increasingly digitised world.

Inclusion

Issues of accessibility, inclusivity, and equity are inherent to any participatory and deliberative process. Proponents argue that digital tools can promote more inclusive decision-making by bringing people together across geographies and time zones, while reducing environmental impact by removing the need for participants to travel (Afzalan and Muller, 2018; Willis et al., 2021). However, digital tools do not guarantee more inclusive outcomes and can lead to new forms of digital exclusion and marginalisation, particularly for those living in rural and remote areas (Pham and Massey, 2018). Socio-economic factors, such as gender, age, race, ethnicity, disability, education, and income, can further exacerbate disparities, and the use of online technologies by governments and/or other powerful actors may empower more affluent communities and further marginalise vulnerable groups (Bricout et al., 2021). Addressing these issues requires considering underlying barriers and inequalities rooted in contextual factors, including technical issues, digital and spatial literacy, and confidence using different tools.

Social and contextual cues

As suggested by Rowe and Gammack (2004, p. 45), one of the most significant features of virtual engagement is its 'attenuation of social context cues that are available in face-to-face interaction' and the reduction of non-verbal cues, such as gestures and facial expressions, that 'form an important part of human communication, and that indicate aspects such as strength of feeling, or whether the other person already knows or understands what is being said' (p. 46). Despite significant technological advances since this paper was published, these concerns remain true and shape how information is transferred, who exchanges information with whom, and what information is communicated. For example, Hafferty (2023) found that it could be more challenging to elicit in-depth and nuanced qualitative information about people's values, opinions, and attachments using solely digital and remote technologies. Digital engagement can be perceived as being more effective at capturing more structured, quantitative information based on specific topics (Hafferty, 2022). Others have criticised digital participatory methods for being perceived by participants (and facilitators) as more structured, awkward, and uncomfortable compared to in-person encounters (Adams-Hutcheson and Longhurst, 2017; Willis et al., 2021). Furthermore, numerous studies reported that the COVID-19 pandemic affected the feasibility of in-situ methods due to reliance on remote technologies for engagement (e.g., Howlett, 2022). In their study of practitioners' perspectives on digital engagement, Hafferty (2022) found that digital tools were perceived to limit the capture and representation of contextual data, which was particularly problematic for planning and environmental decision-making processes which sought to understand place-based issues and incorporate local knowledge.

Power

Critical social scientists have exposed the pitfalls and limitations of participation for promoting practices which glaze over important questions of power, and may actually work to systematically reinforce (rather than overthrow) inequalities (Cooke and Kothari, 2001; Kesby et al., 2007; Stirling, 2008). Although some studies do not explicitly consider how power relations differ in digital engagement, or do not find any different power dynamics between in-person and digital engagement (Afzalan and Muller, 2018; Panchyshyn and Corbett, 2022; Willis et al., 2021), others claim that digital techniques can help empower participants. For example, Marzi (2021) found that digital and remote participatory video methodologies offer new avenues for participants to take control over the research process, while building technical skills and digital literacies that have long-term value to participants (also see Marzi, 2023). Research during the COVID-19 pandemic highlighted the importance of reflexivity and emphasised the need to use digital tools within the context of an explicit reevaluation of roles and relationships within research and practice, in order to minimise top-down hierarchies and concentrated decision-making power (Marzi, 2021).

Trust

Proponents claim that digital tools can enhance trust and transparency by facilitating consensus-building, social bonds, and collective action (Afzalan and Muller, 2018). Digital participatory platforms are often marketed as effective solutions for maximising transparency in decision-making, which can in turn increase the perceived trust and legitimacy of decisions and decision-making institutions (Falco and Kleinhans, 2018). For example, UK-based online citizen engagement platform, Commonplace, highlights the benefits of their participatory tools for ensuring that 'every comment made by a member of the community is visible to the community' in acknowledgement that 'transparency is vital in building trust in the planning system' (Commonplace, 2021, p. 31). However, *trust* is highly subjective and multi-dimensional and there is growing evidence that maintaining trusting relationships with participants can be more challenging in digital and remote environments. Some argue that meeting in-person with participants before any online engagement is conducted is crucial for fostering and maintaining trusting relationships (Hafferty, 2022). Face-to-face meetings are valued for fostering more informal and unstructured environments which support free-flowing conversation and greater human connection, which can be difficult to meaningfully replicate online (Hafferty, 2022).

Concluding remarks

Digital tools are continuously shaping participation in planning and environmental decision-making processes, with bold optimism in the potentials of technology for improving more inclusive and just sustainable transitions. In this review, we have argued that against a backdrop of increasing digitalisation of participatory processes in research and practice, supported by national and international strategies that frame technological innovation as a "win-win" for delivering on socio-economic and environmental goals, there is increasing uncertainty about their effectiveness at delivering more meaningful and inclusive outcomes. Our review has synthesised recent work on the use of digital tools for participatory planning and environmental decision-making in the UK, highlighting complexities and discontents regarding their ethical and wider societal implications. We have considered how contextual factors and socio-economic dynamics – which are fundamental to debates around *any* participatory process, regardless of the methods used – take on new dimensions in remote digital environments, compared to in-person situations. To structure these debates, we focus on four interlinked factors which have been empirically shown to shape outcomes in participatory processes: inclusion, social and contextual cues, power and trust (Hafferty, 2022; Hafferty et al., in press). This chapter has uncovered the complexity of debates surrounding digital participation, highlighting how digital tools may have multiple, interrelated and highly context-dependent challenges.

Despite growing awareness of the threat of technology for just sustainable transitions, there is a need for more empirical evidence of what works, what does not work, how digital tools impact whose voice is heard (and who is left behind), what knowledge is produced and considered legitimate in increasingly digitised spaces, the consequences for the erosion of trust and perceived legitimacy of decisions, among other persistent issues. Our chapter has demonstrated the importance of taking a more critical and ethical approach to the application of digital tools for participatory planning and environmental decision-making. However, it is also important to acknowledge several limitations to our study that offer opportunities for future research. While we have emphasised the significance of methodological choice in the design of participatory processes, the selection of methods alone may not fully explain the disparities between expectations and realities (Wesselink et al., 2011). It is likely that broader socio-economic, cultural, institutional and governance factors play a significant role in explaining engagement outcomes, and future research should investigate the extent to which these factors shape outcomes in entirely digitised processes. Although we draw from research conducted in planning and environmental-decision-making, the issues discussed will likely hold for digital participatory processes in other contexts (e.g., healthcare, technological innovation, and education). Future research should further explore whether and how new considerations, which may include or stretch beyond those highlighted in this review, arise from the increased digitalisation of participatory processes, both within and beyond environmental, planning, and sustainability arenas. We support arguments that there are no purely technological solutions to complex environment and sustainability issues, and call for more cautious and critical approaches rooted in social justice and ethical practice.

Note

1 Hafferty et al., in press.

References

Afzalan, N. and Muller, B. (2018) 'Online participatory technologies: Opportunities and challenges for enriching participatory planning', *Journal of the American Planning Association*, 84(2), pp. 162–177. https://doi.org/10.1080/01944363.2018.1434010.

Babelon, I. (2021) *Digital Participatory Platforms in Urban Planning*. Northumbria University. https://ethos.bl.uk/OrderDetails.do?uin=uk.bl.ethos.823362.

Baker, S. and Chapin, F.S. (2018) 'Going beyond "it depends:" The role of context in shaping participation in natural resource management', *Ecology and Society*, 23(1). https://doi.org/10.5751/ES-09868-230120.

Bricout, J., Baker, P.M.A., Moon, N.W. and Sharma, B. (2021) 'Exploring the smart future of participation: Community, inclusivity, and people with disabilities', *International Journal of E-Planning Research*, 10(2), pp. 94–108. https://doi.org/10.4018/IJEPR.20210401.oa8.

Brown, G., Reed, P. and Raymond, C.M. (2020) 'Mapping place values: 10 lessons from two decades of public participation GIS empirical research', *Applied Geography*, 116, p. 102156. https://doi.org/10.1016/j.apgeog.2020.102156.

Carver, S., Evans, A., Kingston, R. and Turton, I. (2001) 'Public participation, GIS, and cyberdemocracy: Evaluating on-line spatial decision support systems', *Environment and Planning B: Planning and Design*, 28(6), pp. 907–921. https://doi.org/10.1068/b2751t.

Charlton, J., Babelon, I., Watson, R. and Hafferty, C. (2023). Phygitally smarter? A critically pragmatic agenda for smarter engagement in British planning and beyond. *Urban Planning*, 8(2), pp. 17–31.

Commonplace (2021) 'Engaging for the Future', (January), pp. 1–40. Available at: https://www.commonplace.is/hubfs/Engaging for the Future.pdf?hsCtaTracking=f2f7a455-4eac-493b-865b-03678a40faab%7Cd2126c33-2397-4433-afaa-61110da90ed2. (Accessed 10 November 2023).

Conroy, M.M. and Evans-Cowley, J. (2005) 'Informing and interacting: The use of E-government for citizen participation in planning', *Journal of E-Government*, 1(3), pp. 73–92. https://doi.org/10.1300/J399v01n03_05.

Cooke, B. and Kothari, U. (eds.) (2001) *Participation: The New Tyranny?* London: Zed Books.

Delaney, J. (2022) 'Minecraft and playful public participation in urban design', *Urban Planning*, 7(2), pp. 330–342. https://doi.org/10.17645/up.v7i2.5229.

European Commission. (2023a). Towards the next technological transition: Commission presents EU strategy to lead on Web 4.0 and virtual worlds. *European Commission*. Online at: https://digital-strategy.ec.europa.eu/en. (Accessed 10 November 2023).

European Commission. (2023b). Promoting inclusive democracy in the digital age: EU and Denmark launch the Digital Democracy Initiative. *European Commission*. Online at: https://international-partnerships.ec.europa.eu/news-and-events/news/promoting-inclusive-democracy-digital-age-eu-and-denmark-launch-digital-democracy-initiative-2023-03-29_en/. (Accessed 10 November 2023).

Evans-Cowley, J. and Hollander, J. (2010) 'The new generation of public participation: Internet-based participation tools', *Planning Practice and Research*, 25(3), pp. 397–408. https://doi.org/10.1080/02697459.2010.503432.

Falco, E. and Kleinhans, R. (2018) 'Digital participatory platforms for co-production in urban development: A systematic review', *International Journal of E-Planning Research*, 7(3), pp. 1–27. https://doi.org/10.4018/IJEPR.2018070105.

Fox, N. et al. (2022) 'Gamifying decision support systems to promote inclusive and engaged urban resilience planning', *Urban Planning*, 7(2), pp. 239–252. https://doi.org/10.17645/up.v7i2.4987.

Galeote, D.F., Rajanen, M., Rajanen, D., Legaki, N.Z., Langley, D.J. and Hamari, J. (2021) 'Gamification for climate change engagement: Review of corpus and future agenda', *Environmental Research Letters*, 16(6). https://doi.org/10.1088/1748-9326/abec05.

Hafferty, Caitlin (2022). *Engagement in the digital age: Practitioners' perspectives of the challenges and opportunities for planning and environmental decision-making*. PhD thesis: University of Gloucestershire, UK.

Hafferty, C., Babelon, I., Berry, R., Brockett, B. and Hoggett, J. (forthcoming) 'Digital tools for participatory environmental decision-making: Opportunities, challenges, and future directions'. In: Sherren, K., Thondhlana, G., and Jackson-Smith, D. (eds). *Opening Windows: Emerging Perspectives, Practices and Opportunities in Natural Resource Social Sciences*. Denver: Utah State University Press.

Hafferty, C., Reed, M. S., Brockett, B.F.T., Orford, S., Berry, R., Short, C., Davis, J. (in press) Engagement in the Digital Age: Navigating the Technical and Ethical Debates Around Participatory Technologies in Environmental Decision-Making. Available at SSRN: https://ssrn.com/abstract=4626765 or http://dx.doi.org/10.2139/ssrn.4626765

Haklay, M. (2013). 'Citizen science and volunteered geographic information: Overview and typology of participation'. In: Sui, D.Z., Elwood, S., and. Goodchild M.F. (eds). *Crowdsourcing Geographic Knowledge*, pp. 105–122.

Howlett, M. (2022) 'Looking at the "field" through a Zoom lens: Methodological reflections on conducting online research during a global pandemic', *Qualitative Research*, 22(3), pp. 387–402. https://doi.org/10.1177/1468794120985691.

Kahila-Tani, M., Broberg, A., Kyttä, M. and Tyger, T. (2016) 'Let the citizens map—public participation GIS as a planning support system in the Helsinki master plan process', *Planning Practice and Research. Routledge*, 31(2), pp. 195–214. https://doi.org/10.1080/02697459.2015.1104203.

Kahila-Tani, M., Kytta, M. and Geertman, S. (2019) 'Does mapping improve public participation? Exploring the pros and cons of using public participation GIS in urban planning practices', *Landscape and Urban Planning*, 186, pp. 45–55.

Kesby, M., Kindon, S. and Pain, R. (2007) 'Participation as a form of power: Retheorising empowerment and spatialising Participatory Action Research', *Participatory Action Research Approaches and Methods: Connecting People, Participation and Place*, pp. 19–25. https://doi.org/10.4324/9780203933671-14.

Kindon, S. (2009) 'Participation'. In: Smith, S.J., Pain, R., Marston, S., and Jones, J.P. (eds). *SAGE Handbook of Social Geographies*. London: SAGE Publications, pp. 517–545.

Marzi, S. (2021) 'Participatory video from a distance: Co-producing knowledge during the COVID-19 pandemic using smartphones', *Qualitative Research* [Preprint]. https://doi.org/10.1177/14687941211038171.

Nabatchi, T. and Leighninger, M. (2015) *Public Participation for 21st Century Democracy*. Hoboken, NJ: John Wiley & Sons.

Panchyshyn, K. and Corbett, J. (2022) 'Pandemic participation', *International Journal of E-Planning Research*, 11(1), pp. 1–12. https://doi.org/10.4018/ijepr.299547.

Pánek, J. (2016) 'From mental maps to geoparticipation', *The Cartographic Journal*, 53(4), pp. 300–307.pan

Pánek, J. (2018) 'Emotional maps: Participatory crowdsourcing of citizens perceptions of their urban environment', *Cartographic Perspectives*, (91), pp. 17–29.

Pánek, J. (2019) 'Mapping citizens' emotions: participatory planning support system in Olomouc, Czech Republic', *Journal of Maps*, 15(1), pp. 8–12.

Pham, L. and Massey, B. (2018) 'Effective digital participation: differences in rural and urban areas and ways forward'. In: *Digital Participation Through Social Living Labs: Valuing Local Knowledge, Enhancing Engagement*, pp. 315–332. Chandos Publishing. https://doi.org/10.1016/B978-0-08-102059-3.00017-4.

Pickles, J. (Ed.). (1995). *Ground Truth: The Social Implications of Geographic Information Systems*. New York: Guilford Press.

Rawat, P. and Yusuf, J.-E. (Wie) (2019) 'Participatory mapping, e-participation, and e-governance', pp. 147–175. https://doi.org/10.4018/978-1-5225-5412-7.ch007.

Reed, M.S., Vella, S., Challies, E., De Vente, J., Frewer, L., Hohenwallner-Ries, D., Huber, T., Neumann, R.K., Oughton, E.A., Sidoli del Ceno, J. and van Delden, H. (2018) 'A theory of participation: what makes stakeholder and public engagement in environmental management work?' *Restoration Ecology*, 26, pp. S7–S17.

Rowe, G. and Gammack, J.G. (2004) 'Promise and perils of electronic public engagement', *Science and Public Policy*, 31(1), pp. 39–54. https://doi.org/10.3152/147154304781780181.

Stirling, A. (2008) '"Opening up" and "closing down": Power, participation, and pluralism in the social appraisal of technology', *Science Technology and Human Values*, 33(2), pp. 262–294. https://doi.org/10.1177/0162243907311265.

United Nations. (2023). *The 17 Goals. United Nations*. Online at: https://sdgs.un.org/goals. Last accessed: 23/08/2023.

Wesselink, A., Paavola, J., Fritsch, O. and Renn, O. (2011) 'Rationales for public participation in environmental policy and governance: Practitioners' perspectives', *Environment and Planning A*, 43(11), pp. 2688–2704. https://doi.org/10.1068/a44161.

Willis, R., Yuille, A., Bryant, P., McLaren, D. and Markusson, N. (2021) 'Taking deliberative research online: Lessons from four case studies', *Qualitative Research*, p. 146879412110634. https://doi.org/10.1177/14687941211063483.

Wilson, A. and Tewdwr-Jones, M. (2021) *Digital Participatory Planning*. New York: Routledge. https://doi.org/10.4324/9781003190639.

7

DIGITAL FRACTURES

Sustainability and the partiality of climate policy simulation models

Ruth Machen

Introduction

Sustainability has always been about more than just climate. Yet, as the urgency of reducing carbon dioxide and associated emissions grows, discourses and practices of environmental concern in Western societies are increasingly focusing on quantifying, calculating, and modelling carbon and its effect on planetary warming and thermodynamic instability. The rise in digital technologies offers new modalities for making the effects of emissions visible and helps to facilitate the commodification of carbon as a resource in vibrant international trading markets touted as prime vehicles for addressing climate concerns. Whilst for many, climate change is a symptom of a wider 'broken' capitalist relationship with the planet (Newell and Paterson 2010; Aronoff 2021), the urgency of reducing carbon is ushering forth many forms of action, only some of which share the wider concerns of sustainability. With a lack of adequate policy and behavioural responses to the climate challenge generating concern worldwide, digital technologies – with their capabilities for large volume data tracking, rapid data processing and prediction – promise to accelerate decision-making, develop more effective precision interventions, and/or improved decision maker understanding (c.f Bakker and Ritts 2019). Digital models – like the EnRoads model in focus throughout this chapter – aim to visibilise the effects of policy decisions and allow experimentation with alternatives in order to stretch the ambition of decision makers. In the context of this rise in digital policy tools for sustainability, it is important to reflect critically on their effects, both on the processes and outcomes of decision making.

One organisation that is pioneering digital modelling for climate governance is Climate Interactive – a US-based think-tank with a purposely transformative agenda. Climate Interactive is a spin off from the US-based Sustainability Institute

DOI: 10.4324/9781003441311-8

(aka Meadows Institute). It formed in 2008 to apply the techniques of systems dynamics modelling developed at MIT (Sterman 2000) alongside theories of organisational learning (Senge 1990) to the complex problem of climate change. Commitment to sustainability (in its broadest sense) has been a foundational for the non-profit since its formation, and combined commitments to carbon reduction and wider social and environmental concerns such as equity, health, biodiversity, and clean water are shared by both its founding co-directors – it was always about "global sustainability, it wasn't climate" (Interview Co-Director B, 2019). However, as climate became the way that the world began to talk about sustainability, Climate Interactive increasingly focused on digitally simulating the effects of carbon and methane on climate as they sought to connect with, and influence, key spaces of influence such as global COP negotiations. Wider sustainability concerns – described within Climate Interactive as 'multisolving' concerns – formed their own distinct programme within Climate Interactive since 2015. However, in 2022, the decision was taken to split the organisation, with the new Multi-solving Institute led by Co-director A breaking away from Climate Interactive, which continued its focus on carbon mitigation under the leadership of Co-director B.

Drawing from theoretical literature on the metaphor of 'fracture', this chapter explores the tensions in holding together the agendas of climate and multi-solving simultaneously, and specifically the difficulties faced in combining these through digital modelling processes. By examining the tensions, contradictions and break points that arise when digitalisation processes fail to capture social (and wider environmental) sustainability concerns, it is argued that digital sustainability is not a win-win approach but that wins in some areas involve losses in others. Specifically, I show that one of the pitfalls to digital sustainability approaches lies in the difficulties that digital modelling faces in capturing, representing, calculating or drawing relationships with qualitative and socio-political concerns. Despite the promise that digital technologies hold for modelling future temperatures and projecting the impact of hypothetical decisions, they can only work with certain types of data and can only factor certain parameters. As a result, a key pitfall in turning to the digital to advance sustainability is that qualitative sustainability concerns beyond carbon may fall out of focus.

Climate Interactive and their digital simulation models

Climate Interactive designs, builds, and utilises interactive methods of learning with the aim of improving climate change decision-making among policy makers and communities worldwide. Digital simulation models are central to their approach with their first simulation model C-ROADS designed to quantify and adding up all the pledges being made by COP delegates on behalf of their national parties to show the extent to which these individual commitments met (or failed to meet) the collective target of limiting global temperature rise to 2 degrees. Since Copenhagen, the C-ROADS model has been used in many negotiating spaces around the world, and a series of role play interactive games have been developed around the model, making it useful for awareness raising and engagement in community and educational settings.

In 2018, Climate Interactive launched their second digital policy simulation model – En-ROADS which considered the sectoral makeup from which promised emissions might come. In contrast to C-ROADS, which directed attention to the extent of the problem, En-ROADS aims to help decision-makers to develop solutions, prompting them to explore what types of policy decision around energy, transport, buildings, and land use need to be put in place in order to limit temperature rise to the post Paris 2015, 1.5-degree cap. In addition to designing and developing the En-ROADS model, Climate Interactive also operate a programme of outreach and engagement providing free web-based training that aims to build a community of En-ROADS advocates and users around the world who can lead engagements with the model in their own political, community, or educational settings worldwide. They also work with policy makers and corporate boardrooms across the United States and beyond. To date, engagement with En-Roads has been huge, with more than 249,000 registered workshops worldwide engaging 4,372 leaders across government, business, NGOs, and academia around the world (CI, Webinar 20th April 2023).

The En-ROADS interactive user interface shown in Figure 7.1 can be accessed freely online (En-ROADS 2019). It presents users with a range of sliders representing policy decision levers that can be adjusted to reflect policy decisions about forms of economic incentive or regulation that are in place to encourage change in the energy mix, levels of building insulation, the extent of land diverted to carbon offsetting etc. The projected temperature rise is displayed in the top right-hand corner of the interface, providing both a goal and an adjudicator of how sufficient these policy decisions are in reducing the degree of warming. The user's aim is to reduce this figure to the 1.5 degrees warming cap that was agreed at the Paris COP, 2015, or below.

Multi-solving

A key challenge that the organisation has faced has been how to integrate wider sustainability concerns into the En-ROADS digital simulation model. These wider concerns are described by Climate Interactive as 'multisolving' concerns: ways of addressing climate change that also benefit health, social inequality, biodiversity, etc. 'Multisolving' – a term developed by co-director A – is about confronting multiple sustainability concerns simultaneously – where "people pool expertise, funding, and political will to solve multiple problems with a single investment of time and money" (Sawin 2018a). Without leaving behind any of the urgency of tackling climate change it focuses attention on the ways in which climate action can also address wider sustainability concerns. An example might be green infrastructure where:

> greening the urban environment can help moderate heat waves. It can also reduce flooding … helps save energy (and thus money) by keeping buildings

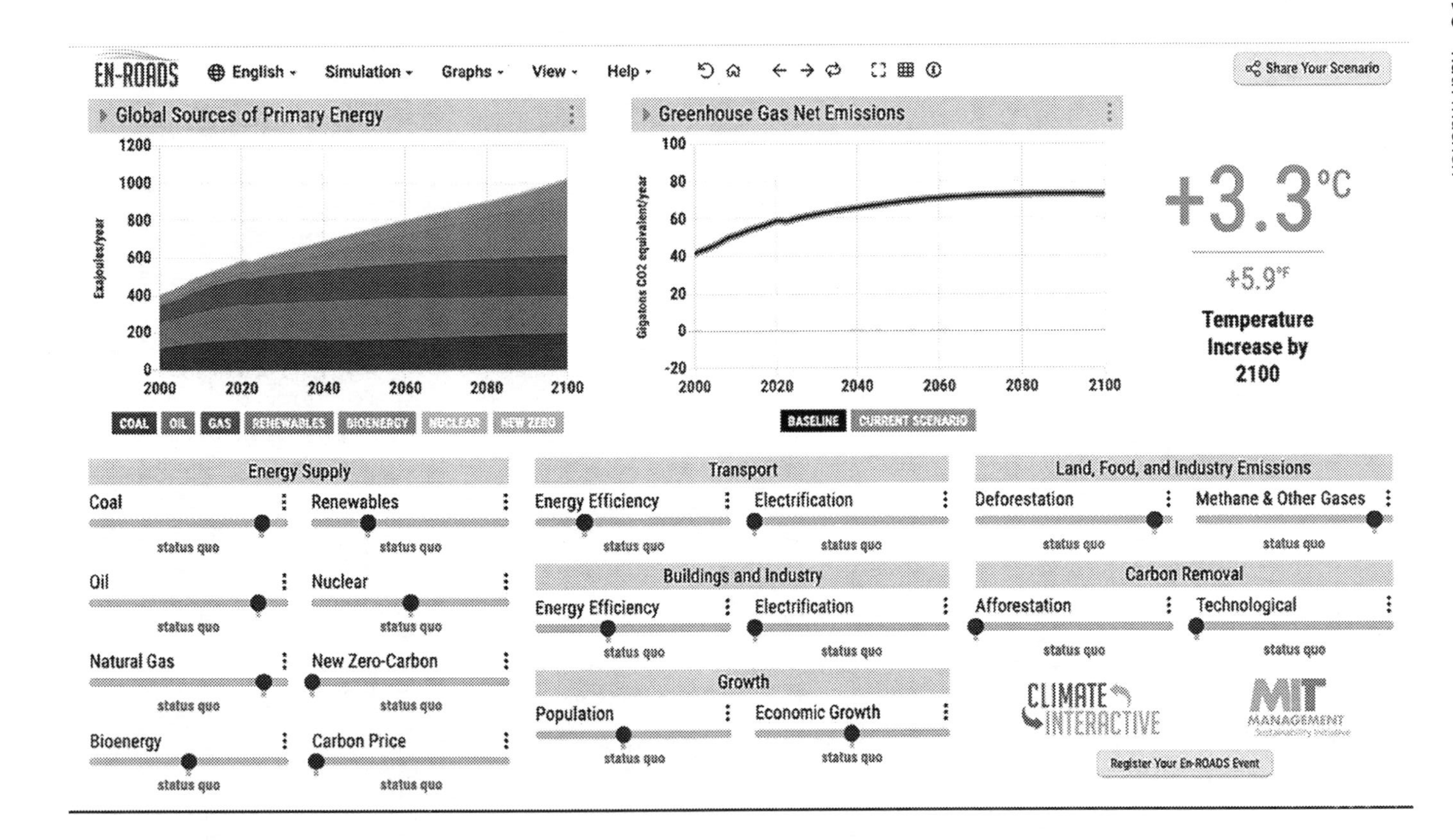

FIGURE 7.1 ENROADS user interface, 2019.

> cool.... improve air and water quality and boost people's sense of well-being. If the green spaces include fruit and nut trees or gardens, they can also help improve access to fresh fruits and vegetables.
>
> *(Sawin 2022)*

Multi-solving rejects notions of having to sacrifice today to protect the climate for tomorrow and instead, champions intragenerational as much as intergenerational equity, looking for positive benefits of climate action today (Sawin 2018b).

In many senses, the tasks of multi-solving are the opposite of computer simulation in that "they need no new apps or state-of-the-art techniques to work" (Sawin 2018a). Instead, they claim a new way of thinking – it is "a way, not a what. It's a way of framing things, it's a way of conceptualising things" (Interview with CI Multisolving team, 2021). Multi-solving draws directly from Donella Meadows observation that "solutions are as interconnected as problems" (Sawin 2023). By looking at how multiple concerns come together in complex and interconnected ways, it tries to overcome siloed thinking and look to qualitative system interconnections that may or may not be easy to quantify, giving people "shared language and tools to communicate these complex ideas" (Interview with CI Multisolving team, 2021). As Co-director A describes:

> multisolving really at its essence is trying to meet the world the way it is ... our minds might find it easier to think in parts but we are in a reality that's an interconnected whole and we are going to have a lot better luck if we start to work together in this way.
>
> *Sawin (2022)*

At the heart of multi-solving is the lived experience of people and a way of operationalising a commitment that everyone matters (Sawin 2018). It involves profound mutual recognition and respect – "to multisolve is to care about someone else's problem as much as your own" (Sawin 2018b). Rather than expressing relationships numerically and causally, multi-solving takes the form of storytelling, one employee described – storytelling even on the smallest scale about how to create a better world and share examples that show another world is possible.

Digital fractures

From a longstanding passion of co-director A in particular, multi-solving became a distinct programme within Climate Interactive in 2015. However, multi-solving faced constant integration problems as one programme officer described –

> for an organisation that talks about the fact that we live in complex systems and shouldn't be working in silos, we found ourselves very much silo'd [laughter].... Even before En-ROADS was released months and months before, we

> had had these discussions about just trying to bring our work together more because it really didn't make sense the way that we were not collaborating.
>
> *(Interview, 2021)*

Despite the commitment to address multi-solving concerns clearly being shared across Climate Interactive, there were difficulties in building these considerations into the En-Roads model itself: "we spent months even on talking about how we would possibly bring multi-solving into this simulation. It was not so much a question of will we....[but] *how* are we going to do this?" (Interview, CI Multisolving team, 2021).

The first problem was one of quantification. Many multi-solving benefits are either difficult to quantify, or simply do not have the data to substantiate cause and effect claims with key variables like carbon, and so are difficult to factor into the differential equations that make up the basis of the model. As a programme lead for En-ROADS described,

> because it's a computer simulation, we're really limited by things that are easily quantifiable. There's a whole space of things that you just can't quantify well. [e.g.] What is the impact of social movements...We can't quantify that... there are only so many things that we can put numbers to, and for it to make sense to put numbers to.
>
> *(Interview, 2021)*

A lack of quantification also made configuring causal modelling statements difficult. With multi-solving concerns

> we couldn't make any statements, you know! En-ROADS is about the statements, 'if this then that', and that's not the way it is with multisolving at all...we can't say this should happen ... you have to say *could*, it *might*, it might increase equity. There are places around the world where this <u>has</u> happened. That was a difficulty.
>
> *(Interview, CI Multisolving team, 2021)*

In this case, the epistemological structures of knowledge-making confined how relationships can be constructed:

> En-ROADS is a very quantitative mathematical rigid thing and so how do we bring in something that is inherently not rigid, and is qualitative, and has to do with storytelling...[bringing] social science and hard science together, that was a big challenge.
>
> *(Interview, CI Multisolving team, 2021)*

The second problem is one of scale. En-ROADS is a global model and "we see multi-solving happening more on a localised grassroots level. So that was another difficulty" (Interview, CI Multisolving team, 2021). This challenge was echoed by one En-ROADS modeller who described the way that "multi-solving benefits are really local. So, it's really challenging to have them in a global model...it's just hard to quantify on the local level in the global model" (Interview, 2021). From the user interface engagement side, the local scale was again highlighted –

> when we face inequity in the world...it's experienced by the individual...with a global model that is focused predominantly on quantifiable things, ...you can lose sight of the people and the experiences...When it comes to policy making, the ways in which we will help resolve global inequity is in the details of the policy, not in the bold slider moves that we capture well with En-ROADS.
>
> *(Interview, CI Engagement team 2021)*

A third challenge lay in the legitimacy of the model itself. It was important for the success of the model to keep a degree of parity in levels of speculation within modelled relationships, such that adding a new variable did not risk undermining the credibility of the whole model. Co-director B illustrated this through the example of including the growing civil society movement: Whilst they could add "unconvinced people, kind of convinced people, very convinced people" and quantify the relationship here between public perception and levels of taxation – for example, "for every unit of belief, how much does the tax on coal go up?" the question, he describes, is twofold – "do we think we can do it well enough, and consistent with the level of confidence of everything else in the model, and will the people who use it have enough confidence?" (Interview 2019). The risk, when modellers have to speculate either about quantified numbers or about the nature of relations between variables he described, is that all those other equations might get called into question, so "it would not be appropriate to add that into this model where we have such confidence and so much of the other structure" (Interview, 2019). Not only could this affect the perceived credibility of other assumptions within the model, it could also affect the benchmarking of the En-ROADS model against other respected models. For Climate Interactive, this is an important way of conferring legitimacy on the model and its results, and an important way that they test model robustness –

> we can compare it to other models in this class, as well. The SSPs, the Energy Modelling Forum models, other Integrated Assessment models; it's comparable to them. We submit them to the IPCC's 1.5 °C Report and we feel good about it being in that class of models. Whereas, if we had parameters around the growth of civil society support, it wouldn't feel as confident.
>
> *(Interview, 2019)*

When the goal is to develop a model that is trusted enough to generate transformative change in decision makers' thinking, there is no scope for doubting any of the model relations, so only the most robust quantifications make the cut.

Finally, a distinctive unique selling point of the En-ROADS model is its speed of calculation and responsivity of results. Processing speed is vital to its success with policymakers and has been a specific area of technical innovation for the organisation. However, when building a model that has to run quickly to provide real-time results, certain compromises have to be made on model complexity – something that Climate Interactive are very open and transparent about. The lead En-ROADS modeller described the way in which modelling as a whole is driven by simplicity, but the En-ROADS model in particular, as a particularly efficient, agile, and 'light' model, needs to capture only parameters which need to be there – either for model results (the most important drivers of system behaviour) or what the organisation term 'face validity (those parameters that don't make much difference to the model itself but need to be there for users to make sense of the sliders or their results). This leanness is vital to enable a model that is computationally light enough to run via a web-based platform and offer instantaneous simulation. The result however is that the modelling process becomes interested only in primary drivers of changes in the model, not nuances. The skill of modelling lies in "being able to weed out the stuff that you don't need, the complexities that aren't going to drive the outcome one way or the other enough to matter" (Interview, En-ROADs modelling team, 2021). Deciding what to include tends to come down to a modelling sensitivity analysis which "helps to identify 'what inputs are the most impactful'" (Interview, En-ROADs modelling team, 2021). Like any model, En-ROADS isn't "capturing all of reality" but only "enough to drive the model efficiently, so that it can be used in real time" (Interview, En-ROADs modelling team, 2021). Multi-solving concerns such as health, equity, biodiversity, wellbeing, and access to clean water, although important for life, are simply outcompeted by other variables like type of energy fuel, or carbon storage options as core drivers of carbon intensity and energy demand in the model.

Thus, whilst Climate Interactive as an organisation remains committed to both carbon mitigation and multi-solving agendas, these challenges about what can be quantified, what can be modelled, what can be positioned as causal, and what scales and types of data this particular global, lean, quantified model can integrate, has meant that "we don't specifically model most of the multi-solving benefits" (Interview, En-ROADs modelling team, 2021). Instead, the organisation adopted several workarounds that prompt multi-solving conversations *around* the model rather than *through* it in order to try to keep multi-solving conversations uppermost.

First, qualitative information about co-benefits was included via information tags within the model "if you look within the simulation when you hit the little 'i' for more information, each slider has co-benefits and equity considerations listed" (Interview, CI Multisolving team, 2021). Second, in many of the facilitated En-ROADs workshops, the facilitator actively prompts conversations around co-benefits. A supplemental slide deck has been developed to provide case studies and narratives that can help illustrate wider multi-solving implications and can easily be drawn upon in pauses during the modelling workshop. Third, a specific

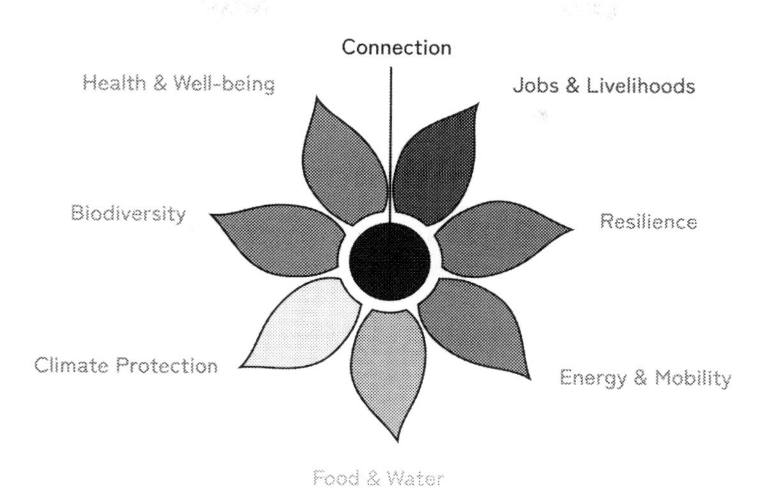

FIGURE 7.2 Multisolving FLOWER tool, see https://www.multisolving.org/flower/.

analogue tool called 'FLOWER' – Framework for Long Term, Whole System, Equity based Reflection – has been designed by the multi-solving programme to help prompt conversations about co-benefits (Figure 7.2):

FLOWER helps identify the benefits or impacts of a policy, investment or action on these seven areas of multi-solving concern, and to prompt discussions over how co-benefits might be improved. It is often used alongside the Enroads model – for example "we did a demonstration for some congressional House representatives … and…they were given this FLOWER tool to go into their groups and talk about: 'If I move this slider, what's an example of a policy that might be applicable to this slider and what are these different co-benefits that are actualised' using FLOWER as a guide" (Interview CI Multisolving team, 2021). In all these ways, multi-solving questions become 'bolted-onto' webinars and workshop sessions around the model and are actively addressed through conversations that take place *around the model and its results*, rather than being explored *through direct experimentation* with the model itself.

The effectiveness of multi-solving questioning is highly dependent on the capabilities and efforts of the workshop facilitators. When sessions are led by Climate Interactive, this is done effectively, yet because multi-solving is not strictly built into the model itself, it does not necessarily travel with the model without Climate Interactive's facilitation. Having worked with planning students to run one such workshop with local planning officers in the North East of England, I experienced the way that, despite our own commitments to multi-solving, it was difficult as inexperienced facilitators to remember to prompt these questions and expand upon the findings of the model in ways that probed the social justice implications in any meaningful way beyond basic questions about who would be disadvantaged by the high pricing of energy. This difficulty in integrating multi-solving within En-ROADS has created frustration that despite these efforts to engage with multi-solving concerns, "people can still run an En-ROADS workshop without ever talking about multi-solving or co-benefits if they want to" (Interview CI Multisolving team, 2021). A member of the multi-solving team described:

> In my mind the ideal would be for it to be impossible, to interact with the En-ROADS simulator without having… some kind of prompt of equity, multi-solving and co-benefits. But as it is right now, people can pick and choose how they want to use the simulator and what stories they want to tell.
>
> *(Interview CI Multisolving team, 2021)*

She goes on,

> I would love to see, if you move a slider, like, you get a pop up box that says: "These are the co-benefits," I would love to see … as you move different sliders, if you could even see a flower diagram and see different petals, shaded and unshaded, and be asked: "Are you going to do this, build new coal infrastructure, …a toggle of": "Are you going to do this with equity in mind or not?" but those things are really hard to put into the equations, right?.
>
> *(Interview CI Multisolving team, 2021)*

Whilst team members across the organisation share a vision for integration in principle, it is the "software capabilities or coding capabilities" (Interview CI Multisolving team, 2021) that prevent this goal being realised.

These experience of the difficulties of working with multi-solving concerns *through* En-ROADS signal the pitfalls of trying to address sustainability concerns through digital models. Despite the dedicated efforts of Climate Interactive to prompt multi-solving concerns, these have to date remained outside the digital and mathematical relations of the model itself. Ultimately, digital solutionism (Kuntsman and Rattle 2019) overlooks the challenges of representing, constructing and framing the less quantifiable values that we ascribe to wider holistic concerns of sustainability beyond carbon. In 2022 Co-director A decided to grow the multi-solving programme outside of the Climate Interactive umbrella, creating an organisational split with the newly formed Multisolving Institute (which focuses on social negotiation), separating from Climate Interactive (which continues its focus on digital simulation models). Although amicable, this organisational fracture is interesting, not so much in management terms, but instead for what it indicates for the concerns of sustainability in a digital world.

Fracture as rearticulation

Recent critical social science scholarship has turned to concepts around breakage – data broken-ness (Pink et al. 2018), fracture (Giglioli and Swyngedouw

2008), fragments (McFarlane 2021) and rupture (Plazky Miller 2021) – to understand contemporary political infrastructural and digital relations. Whilst for some the language of fracture might conjure ideas of breakage in something that was once whole, for feminist science scholars – taking a lead from Donna J. Haraway – there never was such a whole. Instead, any intimation towards a whole is "always constructed and stitched together imperfectly, and therefore able to join with another" (Haraway 1988: 586). As Pink et al. describe, these ideas of partiality and breakage are part and parcel of an academic lexicon for assemblage and entanglement – without one the possibilities of recombination and repair are distant. The process of disarticulation is essential to the possibilities for rearticulation (Laclau and Mouffe 1985). Instead, the concept metaphor of 'broken data' account for the ways that "data might be in processes of decay, making, repair, re-making and growth" (Pink et al. 2018:1). For political ecologists, 'fracture' refers to the unfinished (Giglioli and Swyngedouw 2008:401) and geo-humanities scholars have drawn on the language of 'fractals' and a 'fractiverse' (Law 2015, Lekan 2014, Englemann et al. 2021) to account for multiple realities and parallel conversations between things that do not quite touch. Fractals intimate towards the tensions of holding difference together in ways that do not ever meet.

The concept of fracture is an interesting conceptual device to understand what is going on in Climate Interactive around multi-solving for three reasons. First it provides a metaphor for a condition in which two concerns and data types – quantifiable carbon reduction and qualitative multi-solving benefits – are held in parallel but are never quite able to quite touch. Data itself is fractured, not in a way that breaks an imaginary whole, but in a way that never quite holds together. Here, being fractured is a condition of the juxtaposition of unintegrate-able knowledges. These tensions between quantifiable carbon modelling and wider sustainability concerns were not new for Climate Interactive, and did not form simple lines through the organisation, but rather developed as a fractured condition for all – their attempts to hold together these diverse concerns, a constant work of articulation. Borrowing the term 'fractals' from Lekan (2014) and later Engelmann (2021), multi-solving stories can be seen as fractals – "aesthetic and ontological devices that offer a powerful critique of whole-earth images …[in which] partiality and specificity does not preclude similarity and interconnection" (Engelmann 2021:245). Multi-solving stories give insights into different worlds, not pieces of a jigsaw that will align into a whole. In psychology a fractured identity is a coping strategy to work through incongruence – the product of things that can't be held cohesively together. The condition of fracture within Climate Interactive illuminates these disjunctures and tensions that belie the ongoing work of holding a narrative of the whole together.

Second fracture draws attention to the process of fracturing itself - to the tensions and contradictions that are resolvable only through breakage. This highlights the process of disarticulation by which new possibilities for rearticulation emerge. On the one hand, the tension between two sides grows until it can no longer be sustained and breaks down – it is unsustainable in the literal sense. On the other, fracturing allows the birth of something new. Platzky Millar (2021:10) describes the way in which a rupture "facilitate the emergence of new practices and understandings of the world. Ruptures thus create conditions of possibility for people to explore new social relations and ideas" – they allow for "something else can grow through the cracks" (Plazky Miller 2021:10). Fracture becomes a metaphor for healing that preserves the ability to "sprout anew" (Lopez 2020:114). Instead of seeing fracture as a negative event for Climate Interactive, fracture is a generative opening – "it is the simultaneity of breakdowns that cracks the matrices of domination and opens geometric possibilities" (Haraway 1991:175). The process of fracture is the generative rearticulation that leads to the birth of the new.

So far, fracture and rupture have been used interchangeably. However, whilst similar, each concept embraces subtle differences. For engineers, these differences lie in the material form (hard versus soft, respectively) and in medicine, fracture – as in break – can be partial, whilst rupture – as in disruption – is total (Addas et al. 2018). Here I add to these differences by suggesting that a fracture, unlike a rupture, does not destroy the former entity, its mission instead is to resolve tension in ways that allow co-existence, rather than destructive transformation. It is for this reason that the term fracture is preferred to account for the incomplete break, the continued proximity and interaction of Climate Interactive and the Multisolving Institute which are woven together in a shared mission via a resolution in which one does not disrupt the functioning of the other. As Lopez describe – fracture "evokes the experience of something breaking while holding" (2020:115).

This story of organisational fracture has focussed only on the tensions between carbon modelling and the qualitative local experiences of multi-solving, however, it would be remiss to leave out the pull factors that likely also played a role in the establishment of the Multisolving Institute. Both Climate Interactive directors have always been astute to opportunities, and undoubtedly the global pandemic provided a moment in which people were increasingly questioning the organising principles of our world and ask more fundamental questions about the way we live, work and structure our lives. This may have been seen as an opportunity to connect the dots between escalating climate impacts, biodiversity loss and structural inequity. Sawin describes that "Invitations to write, speak, and teach about multi-solving came fast and furious and with it the possibility [to]… help support leaders around the world to respond to crises with multisolving" (Multisolving Institute website 2023). In this sense, the multi-solving programme perceived that it had found its moment.

If we see fracture not as a break from an originary whole but rather as a process of disarticulation and rearticulation (Laclau and Mouffe 1985) that enable processes of entangling with others and "facilitate the emergence of new practices and understandings of the world" (Platzsky 2021:10), then the fracture becomes a space of growth and new possibility. For Lopez, the language of "fracture" – emerging in relation to Deleuze and Guatarri's work on rhizomes – offers a metaphor for healing, fractures "create new contexts...can facilitate flow and growth in multiple directions and make other connections, fractures, and blockages possible" (Lopezp115). For Co-director A (Director of the new Multisolving Institute), this is an opportunity for rearticulation. "I think of multisolving as repairing some of the fractures that empire and white supremacy and patriarchy and economic extraction, environmental extraction have created in our worlds and every multi-solving project is like a tiny suture in those breakages" (Sawin, 2022). The fracture of multi-solving represents an epistemological shift in which particular knowledges and ways of thinking that could not be contained within the digital approaches of Climate Interactive's practices – in doing so, it shows us that other ways of thinking and working are possible.

Digital fractures and sustainability

Pink et al. argue that if we are to "better understand data futures, we need to account more fully for the incomplete, contingent and fractured character of digital data" (Pink 2018:1). Since the United Nations Conference on the Human Environment 1972 sustainability, expressed through the discourse of sustainable development, has bypassed or 'seamed over' the tensions between environmental impact and economic growth with its unequal, extractivist and polluting relations that underpin multi-solving concerns. Within sustainability's holistic collection of concerns lie multiple tensions and lines of fracture that are not easily explored through an epistemological stance such as digital modelling that posits a singular reality, quantifiable parameters, clear drivers, and a universalisable experience.

The concept of fracture helps to focus attention the limits to this articulation, and by examining these limits we can also better understand the limits of the digital in relation to sustainability. As Plazky Miller (drawing on Alexander) describes, ruptures can be useful in "demystifying and revealing the nature of the structures with which they break (Alexander 2013, 607)" (2021:12). The fracture around multi-solving tells us something about the digital – about what it normalises, what it makes visible and what it obscures. It reveals the occlusions of the qualitative, the local, and the dominant ideologies of abstraction and universalisation that lie behind algorithmic practices (Machen and Nost 2021). The case of Climate Interactive and its struggles to make digital modelling processes capture social (and wider environmental) sustainability concerns lead us to question the assumption (since the Brundtland Report (WCED 1987)) that digital technology is central to driving a vision of sustainability. For the digital modelling devices that offer such resonance for modelling climate (with their propensity to carve up, quantify, join, and predict carbon effects) are limiting articulation of the full range of sustainability concerns. These challenges experienced by Climate Interactive are an important source of learning for attempts to explore sustainability through digital technologies, one that adds to repeated questioning of digital technologies as a panacea by scholars that express concern around the environmental impact of digital processing (c.f. Siddik et al. 2021) or concerns about social equity and systematic bias in algorithmic decision making (c.f. Noble 2018).

Despite their illusion of comprehensivity, models are always partial. Climate Interactive themselves are comfortable with the limits to modelling, recognising that "models can reflect a slice of reality. They never capture all of reality" (Interview, 2021). Yet, when models are used to explore possibilities and potentialities in the world, it is all too easy to forget the ways that this partiality limits the types of conversation that can be held – at the very same time as models offer their own productive potentialities. The EnRoads model does what it sets out to achieve very well – prompting reflection on possible pathways to carbon reduction and challenge over the scope of ambition. This has real value. The fact that carbon is amenable to forms of quantification which resonate with digital modelling techniques may –although we critique it – be a saving grace for progress on carbon mitigation efforts. However, with sustainability referring to more than just carbon but also to the construction of liveable, desirable futures which attend to the preservation of ecological life-support systems and to the promotion of inter- and intra-generational equity – let us not forget those aspects of sustainability that cannot be quantified and expressed in robust mathematical relations. As digital technologies are suited to speeding up and scaling up some forms of knowledge over others, focusing just on carbon and ignoring social, political and wider environmental concerns, fails sustainability and worse could potentially exacerbate both global environmental and social justice issues (Abram et al. 2020).

Conclusion

This chapter has examined the paradox within digitalisation and sustainability: that the more sustainability is explored through digital modelling, the less able it seems to encompass the breadth and interconnectedness of sustainability concerns. Rather than focusing on developing recommendations for how digitalisation and sustainability might better interact it, has instead sought to trace their discontinuities. The story of fracture that this chapter tells is one of more than simple organisational division. Instead, it is a story about the tensions within digital sustainability itself and the difficulties of holding qualitative situated experience and quantifiable 'universalisable' data-based knowledge together. The concept of 'fracture' has been useful for highlighting both the conditions of multiplicity and the processes of disarticulation and rearticulation that "open spaces in which continuities and discontinuities across time, space and experience can be interrogated" (Moore 2004 cited in Pink 2018:4). If digital sustainability is an epistemology – "an array of methods and practices for producing, validating, and applying knowledge to address climate challenges, digital innovations, biodiversity, and their interwoven dynamics" (Iapaolo et al. this volume, p3) – then this epistemology needs to be explicit about its blind spots, its silences, its gaps in datapoints, and in its quantified relations/impacts. STS scholarship has taught us that particular practices bring particular realities into being. Whilst carbon modelling may help bring about much needed carbon reduction, other practices may be required to bring into being wider holistic benefits. Further research examining the knowledge practices of the new Multi-solving Institute (established after this research was conducted) would generate greater insight into what these practices look like and might help establish what role, if any, digitalisation could play in support of these wider ambitions.

The example of Climate Interactive – a small not for profit – is *not* an example of concentrating power in big tech companies. *Nor* is it an example of digital capitalism – extracting ever more combinations of personal data to be traded – often without our knowledge let alone our consent. Nevertheless, *even with the most earnest intent* to address sustainability concerns in their widest sense, the challenges of working through the infrastructures of digital modelling software, the differential calculus (that makes up the model itself), and the web programming code have limited the possibilities for engaging with the full range of sustainability concerns. This reveals limits to digital abilities to engage with qualitative sustainability concerns, and ultimately limits the possibilities for sustainability itself. If the digital is the new battleground where the struggle for a more sustainable, democratic, and inclusive society is fought (Certomà 2021), then we need to ensure this partial vision of that which is quantifiable and calculable does not fail the wider sustainability concerns that it promises to address, and instead remain attentive to where these concerns are being squeezed out of the visions of sustainability that emerge; however unintentionally.

References

Abram, S., Atkins, E., Dietzel, A., Hammond, M., Jenkins, K., Kiamba, L., Kirshner, J., Kreienkamp, J., Pegram, T., & Vining, B. (2020). COP26 universities network briefing / OCTOBER 2020. *Just Transition: Pathways to Socially Inclusive Decarbonisation.* Available online at https://www.ucl.ac.uk/public-policy/sites/public-policy/files/cop26_just_transition_policy_paper_-_final_.pdf (last accessed August 2023).

Addas, F., Yan, S., Hadjipavlou, M., Gonsalves, M., & Sabbagh, S. (2018). Testicular rupture or testicular fracture? A case report and literature review. *Case Report Urology* 1323780.

Aronoff, K. (2021). *Overheated: How Capitalism Broke the Planet--And How We Fight Back.* New York: Bold Type Books.

Bakker, K., & Ritts, M. (2018). Smart earth: A meta-review and implications for environmental governance. *Global Environmental Change*, 52, 201–211.

Brundtland, G.H. (1987). *Our Common Future: Report of the World Commission on Environment and Development.* Geneva: UN-Document A/42/427.

Certomà, C. (2021). *Digital Social Innovation. Spatial Imaginaries and Technological Resistances in Urban Governance.* New York: Palgrave Macmillan.

ENROADS. (2019). *ENROADS User Interface.* Available online at https://en-roads.climateinteractive.org/scenario.html?v=23.6.1.

Engelmann, S., Dyer, S., Malcolm, L., & Powers, D. (2022). Open-weather: Speculative-feminist propositions for planetary images in an era of climate crisis. *Geoforum*, 137, 237–247.

Giglioli, I., & Swyngedouw, E. (2008). Let's drink to the great thirst! Water and the politics of fractured techno-natures in Sicily. *International Journal of Urban and Regional Research*, 32(2), 392–414.

Haraway, D. (1988). Situated knowledges: The science question in feminism and the privilege of partial perspective. *Feminist Studies*, 14(3), 575–599.

Haraway, D.J. (1991). A cyborg manifesto. simians, cyborgs, and women. In D.J Harraway (ed.). *Simians, Cyborgs and Women: The Reinvention of Nature*, (pp. 149–181). New York: Routledge.

Iapaolo, F., Certomà, C., & Martellozzo, F. (2024). Digital (Un)sustainabilities: An Introduction, in "*Digital Technologies for Sustainable Futures:Promises and Pitfall*", (pp. 1–13). Routledge.

Kuntsman, A., & Rattle, I. (2019). Towards a paradigmatic shift in sustainability studies: A systematic review of peer reviewed literature and future agenda setting to consider environmental (Un) sustainability of digital communication. *Environmental Communication*, 13(5), 567–581.

Laclau, E., & Mouffe, C. (1985). *Hegemony and Socialist Strategy: Towards a Radical Democratic Politics*. London: Verso Books.

Law, J. (2015). What's wrong with a one-world world? *Distinktion: Scandinavian Journal of Social Theory*, 16(1), 126–139.

Lekan, T.M. (2014). Fractal earth: Visualizing the global environment in the Anthropocene. *Environmental Humanities*, 5(1), 171–201.

Lopez, J. (2020). Healing is rhizomatic: A conceptual framework and tool. *Genealogy*, 4(4), 115.

Machen, R., & Nost, E. (2021). Thinking algorithmically: The making of hegemonic knowledge in climate governance. *Transactions of the Institute of British Geographers*, 46(3), 555–569.

McFarlane, C. (2021). *Fragments of the City: Making and Remaking Urban Worlds*. University of California Press.

Noble, S.U. (2018). Algorithms of oppression: *How search engines reinforce racism*. New York: New York University Press.

Newell, P., & Paterson, M. (2010). *Climate Capitalism: Global Warming and the Transformation of the Global Economy*. Cambridge University Press.

Platzky Miller, J. (2021). A Fanonian theory of rupture: From Algerian decolonization to student movements in South Africa and Brazil, *Critical African Studies*, 13(1), 10–28.

Sawin, E. (2018a). The magic of "multisolving." *Stanford Social Innovation Review*. https://doi.org/10.48558/W5D4-6430.

Sawin, E. (2018b). *TEDx Talks the Power of Multisolving for People and Climate*. Available online at https://www.youtube.com/watch?v=prF8trTallQ.

Sawin, E. (2022). Op-Ed: We must adapt to climate change. Can we do it in ways that solve other problems too? In *Environmental Health News*. Available online at https://www.ehn.org/climate-change-adaptation-2656805797/weaning-off-fossil-fuels.

Sawin, E. (2022). Systems change, multisolving, and the power to change direction with Dr. Elizabeth Sawin. *Crowdsourcing Sustainability You Tube Videos*. Available online at https://www.youtube.com/watch?v=ge1_pmeck-g.

Sawin, E. (2023). *Multisolving: Taking Action on Climate. Health Equity and Biodiversity as Interconnected Issues*. https://www.youtube.com/watch?v=eAEeQ-rFWDk.
Siddik, M.A.B., Shehabi, A., & Marston, L. (2021). The environmental footprint of data centers in the United States. *Environmental Research Letters*, 16(6), 064017.
Sterman, J. (2000). *Business Dynamics: Systems Thinking and Modelling for a Complex World* (p. 982). Boston: Irwin/McGraw-Hill.
Senge, P. (1990). *The Fifth Discipline – Art and Practice of the Learning Organisation*. Random House Business Books.

PART 2

Twin transition on the ground

Local experimentations with digital sustainability

8

SHARE AN IDEA

AI-augmented urban narrative

Mark Dyer, Shaoqun Wu, and Min-Hsien Weng

Introduction

Public data, contributed by citizens, stakeholders and other affected parties, are increasingly being used to collect the shared ideas of a wider community. Online surveys with questions and answers powered by e-participation tools are being used to identify the public's new ideas, preferences and opinions through public consultation exercises. For example, "The Quality of Life Survey" (Nielsen 2018) undertaken recently in New Zealand asked more than 7,000 residents for their opinions on a wide range of different aspects of urban life in four major cities (Tambouris et al. 2013). Likewise, social media, such as Twitter, is another data source to observe public response to news events, such as an earthquake or presidential election, and collect public opinions (Enli 2017; Sakaki et al. 2010; Wang et al. 2012). Once ideas are collected from public data, the policy formulation process is undertaken to map the individual ideas of self-interests to the shared values of public interests, and then translate into action plans (Dalton et al. 2020: 243; Fischer 2003; Sam 2003). However, large format data from public consultation exercises are often organised and analysed manually for reporting to relevant parties or published online for public access.

This was the case for the 'Share an Idea' public consultation exercise undertaken by the Christchurch City Council after the 2011 earthquake. The public consultation was carried out in response to the New Zealand Government's Canterbury Earthquake Recovery Act 2011 (CER Act) that required Christchurch City Council (CCC) to develop a draft recovery plan for the central city, in nine months (including public consultation), whilst Canterbury Earthquake Recovery Authority (CERA) separately developed a recovery strategy for the Christchurch metropolitan

DOI: 10.4324/9781003441311-10

area. The 'Share an Idea' website generated 58,000 hits and engaged the public in four key areas: move (transportation), market (business), space (public place and recreation) and life (mixed uses). The campaign also used traditional and other social media networks to gather public contributions. The unprecedented level of public participation generated 106,000 ideas over six weeks and these informed the development of the Draft CCP (Brand and Nicholson 2016). Of which, 2,740 quotes (58,100 words) from the original texts were published in the Christchurch Common Themes (CCT) report (Christchurch City Council 2011), which has been used in this study.

The public engagement initiative showed that online public consultation was readily supported by the community, albeit under emergency conditions following the devastating earthquakes. The grassroots process proposed several important design preferences for the rebuild such as limiting the building height in the Central City to six stories. However as explained by Brand and Nicholson (2016), many of the grassroots ideas fell afoul of top-down central government bureaucracy. In this case, CERA took complete control for the Recovery Plan through a top down exercise conducted behind closed doors (Gjerde 2017). The bureaucratic process attracted considerable criticism with subsequent research revealing that CERA was both incapable and unwilling to engage effectively with the public (Simons 2016). Not surprisingly, public perceptions of the way the recovery was managed were perceived as being very poor, with up to 80% of the people living in Christchurch holding a negative view (Simons 2016). Reasons for this are many and are likely to reference a person's experiences of dealing with their own circumstances. So, what went wrong and what lessons can be learnt from the otherwise leading example of public engagement for reconstruction after a series of major earthquakes.

In marked contrast to the relatively short-term strategy deployed by New Zealand government's CER Act for public consultation, the longer-term strategies adopted by the Scandinavian cities of Växjö and Sønderborg emphasised the importance of ongoing public consultation and engagement to reinforce community narratives based on core values that underpinned one of the pillars of sustainable development (Dyer and Ögmundardóttir 2018). In particular, ethnographic studies undertaken for the transition of Scandinavian cities of Växjö and Sønderborg towards becoming fossil-fuel free economies highlight the importance of identifying common narratives to transcend political cycles (Dyer and Ögmundardóttir 2018). In the case of Växjö, the common narrative was one of protecting the environment and restoring damaged ecosystems. In comparison, the successful transition for Sønderborg was built on a narrative of combining job creation with movement towards zero carbon emissions based on renewable energy, deployment of district heating systems and generation, retrofitting of homes, all of which was funded by a public-private partnership between local government, industry and banks. In summary, the lessons learnt from the transformation of cities towards becoming fossil fuel free societies points to the need for 'bottom-up' conversations based on shared cultural values and interests that motivate change.

Elements of successful public engagement

Having recognised the importance of public engagement, it is critical to understand the elements that led to successful engagement. The seminal works by Fung (2003, 2006) characterises public participation as having three distinct elements (or dimensions), namely who are the participants, how the participants communicate and what is the impact of the participation exercise. The approach is developed further as a three dimensional 'Democracy Cube'. In a similar vein, the later work by Nabatchi (2012) advocated a framework for designing public participation comprising eight elements (or propositions). The main characteristics of both frameworks for public participation can be summarised as follows:

- Deliberative modes of communication that avoid one-way communication
- Collaborative processes focussed on common interests (values) as opposed to fixed positions
- Shared decision making to resolve values-based policy conflict
- Provision of information to better inform participants and aid good quality decision making
- Recruitment strategies that are representative of diverse stakeholders and avoid bias

Yet as recognised by Fung (2003, 2006), Nabatchi (2012), Gleeson and Dyer (2017), one of the major challenges with participatory mechanisms is the creation of collaborative two-way processes for representative groups of participants to critically define and identify possible solutions whether it be in the public policy arena or design of a new product. Traditionally the scale of the deliberative process is a controlling factor. As observed by Nabatchi (loc cit), large format processes typically take place in town hall style meetings that tend to foster one-way communication.

This is where artificial intelligence (AI) based on natural language processing (NLP) can provide rapid analysis of public consultation to communicate real time narratives based on shared 'interests' based on shared cultural values to help formulate public policy and design of new services, products and infrastructure.

Coincidentally, the study by Dyer et al. (2017) explored similar issues when attempting to implement a Triple Zero waste strategy for NATO military camps. Again, the study identified poor engagement by military personnel as the principal obstacles to reducing waste generation, water and energy consumption at military operations, not a lack of technology. The study offered a practical framework, as set out below, for implementing a Triple Zero approach by raising awareness and engagement of participants in the following areas. However the NATO study did not provide a means of using the latest AI technology to process large format data into telling or retelling stories to motivate change, which is the purpose of this chapter.

a Create Identities: By creating identities of sustainability, where all of the elements involving society, environment, and economy are present and intertwined, communities begin to recognise sustainability as a concept that affects everyone's lives, families, children, future and happiness and which needs to be communicated in simple language that everyone can understand.
b Tell Stories: By listening, telling and retelling stories about how change can happen and has successfully happened, shared narratives can become powerful human motivators for action. There are countless examples about how communities have succeeded in making a positive change, by reducing carbon consumption and climate gas emissions when dealing more imaginatively with energy, waste and water or other issues that protect the environment in a wider sense.
c Empower Individuals: Empowering individuals who already feel passionate and committed to environmental sustainability. These champions are invaluable co-workers in making change happen, who are often having a hard time being heard and obtaining support for their efforts.

Hence, without adequate sharing of authority, the public are potentially excluded from the design process and all the knowledge and insight that they can bring. As mentioned earlier, the renowned political scientist Dewey (1981) remarked that 'the man who wears the shoe, not the shoemaker, knows best where it pinches'. This insightful comment goes to the heart of participatory design. To be successful, it requires sharing of authority as illustrated in Figure 8.1 and recognition that a design process is iterative. In practice, this means devising feedback systems to facilitate a two-way dialogue that is valid, respectful, and representative of the community whilst creating a less static and more dynamic environment to trial prototypes before selecting more permanent solutions.

AI-aided public consultation

Taking inspiration from the success of Växjö and Sønderborg transitions towards becoming fossil-fuel free economies, the authors investigated the use of NLP tools to compose 'common narratives' from large format data sets such as the Christchurch' 'Share an Idea' initiative. At the same time, it was recognised that there was the added opportunity to align large-scale public engagement with participatory design techniques to address major societal challenges by co-designing new services, products and infrastructure for CE transition that respond to 'user needs'. With these ambitious goals in mind, a digital platform called 'Urban Narrative'

	INFORM	CONSULT	INVOLVE	COLLABORATE	EMPOWER
PUBLIC PARTICIPATION GOAL	To provide the public with balanced and objective information to assist them in understanding the problems, alternatives and/or solutions.	To obtain public feedback on analysis, alternatives and/or decision.	To work directly with the public throughout the process to ensure that public issues and concerns are consistently understood and considered.	To partner with the public in each aspect of the decision including the development of alternatives and the identification of the preferred solution	To place final decision-making in the hands of the public
PROMISE TO THE PUBLIC	We will keep you informed.	We will keep you informed, listen to and acknowledge concerns and provide feedback on how public input influenced the decision.	We will work with you to ensure that your concerns and issues are directly reflected in the alternatives developed and provide feedback on how public input influenced the decision	We will look to you for direct advice and innovation in formulating solutions and incorporate your advice and recommendations into the decisions to the maximum extent possible	We will implement what you decide.
EXAMPLE TOOLS	• Fact sheets • Websites • Open houses	• Public comment • Focus groups • Surveys • Public meetings	• Workshops • Deliberate polling	• Citizen Advisory • Committees • Consensus building • Participatory • Decision-making	• Citizen juries • Ballots • Delegated • Decisions

FIGURE 8.1 International Association for Public Participation (IAP2) Spectrum of Public Participation.

was developed using Stanford NLP toolkits. The platform extracts key messages that contribute towards an overall narrative about shared interests. The concept has been demonstrated using published data from Christchurch 'Share an Idea'.

At the same time, the convergence of public participation techniques with participatory design methods offers the prospects of upscaling public consultation as a two way process to create common narratives based on 'shared interests' using NLP tools (Eisenstein 2019). Examples already exist of NLP being used on popular social network platforms to analyse comments on products, movie reviews or restaurant ratings for marketing purposes. More recently, NLP has made rapid advances using deep learning techniques to develop automatic learning procedures (Hinton, Osindero, and Teh 2006; Manning 2015) and in particular neutral network including word embeddings to capture semantic properties of words or semantic role labelling (such as an agent, goal, or result) to words or phrases to find the meaning of the sentence (Gildea and Jurafsky 2002; Kalchbrenner and Blunsom 2013).

For this study, Stanford NLP Core toolkits (Manning et al. 2014) were employed for several reasons. Firstly, the individual modules were developed using neural language models that gave reliable language analysis results. Secondly, the toolkits use common and uniform Application Programming Interfaces (APIs) that reduce the workloads of system integration. Lastly, central language processing analyses are supported by the toolkits that both identify phrases from a sentence and capture compositional structure of phrases, e.g. the parser extracts the noun phrase "buildings that are accessible to all", in addition to a single noun word "buildings", to give more sensible meanings (Socher, Bauer, Manning, and Ng 2013).

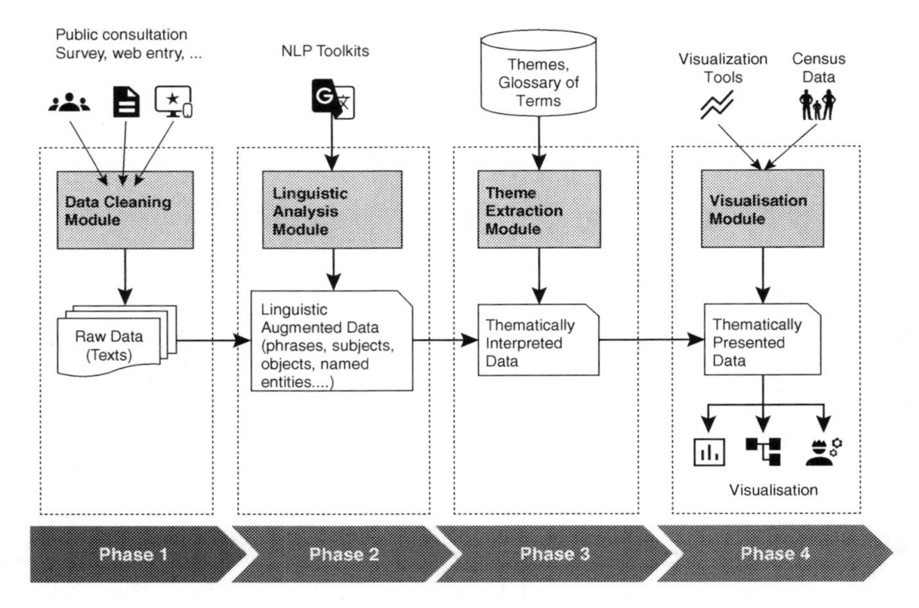

FIGURE 8.2 Systems Architecture for 'Urban Narrative'.

Methodology and results

General

With the aim of creating data stories from NLP analysis of large public format data, a digital platform 'Urban Narrative' was developed using systems architecture comprising 4 phases as illustrated in Figure 8.2.

Phase 1 – Data storage and cleaning

Phase 1 extracts text from various file types (e.g. word documents, pdfs and web pages) and encodes each public quote as a single text, and all the texts are collectively stored as a dataset in a repository or Cloud storage. Since the raw texts may well contain irregularities that would cause NLP toolkits to produce inaccurate result, a series of text cleaning and corrections are conducted that include:

- Replacing tabs and duplicated spaces with a single white space, to ensure every word is separated by one space.
- Replacing abnormal quote marks, such as back quote and double quotes, with the regular quote mark, to ensure every mark remains the same across the entire text.
- Replacing hyphen (-) with commas to unify the symbols.

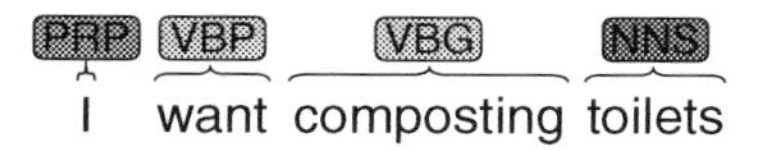

FIGURE 8.3 Part-Of-Speech tagged sentence Credit: Stanford CoreNLP via stanfordnlp.github.io/CoreNLP/.

Phase 2 – Linguistic augmentation

Phase 2 applies Stanford NLP Core toolkit to parse the individual sentences into structured linguistic data (Manning et al. 2014), where each sentence is divided into words and each word tagged with part-of-speech label to classify the word as a noun, verb, object etc. In particular, the process involves three steps. Step 1 splits the texts into individual sentences. The text "I want composting toilets in most homes so the sewerage will not be a problem in the next earthquake." becomes two sentences "I want composting toilets." and "So the sewerage will not be a problem in the next earthquake." Step 2 tokenises each sentence and assigns Part-of-Speech (POS) tags to words in a sentence. Figure 8.3 shows the tagged version of the sentence "I want composting toilets." with POS tags at the top where 'PRP' stands for personal pronoun, likewise 'VBP' for verb, 'VBG' for gerund verb that acts like noun and 'NNS' for plural noun.

Subsequently, Step 3 extracts four types of linguistic data as follows.

1 *Verbs and Nouns* (e.g. *want* as a verb and *toilets* as a noun respectively) are identified based on the POS tags associated with each word.
2 *Noun and verb phrases* are extracted using Stanford Constituency parser (Socher, Bauer, Manning, and Ng 2013; Socher et al. 2013).

   A noun phrase must have at least one central noun word (e.g. *toilets*) that is sometimes modified by other words (e.g. *composting*). Two or more noun phrases can be merged into a single noun phrase. The following sentence contains two noun phrases "recycling bins" and "more trees". To capture a broad semantic meaning, "recycling bins and more trees" is treated as one noun phrase.

   "I want recycling bins and more trees."

   A verb phrase constitutes a main verb (e.g. *want*) followed by noun phrases (e.g. *green spaces*) or sentence segments. For example, the sentence segment "to see more rainwater reused for irrigating green spaces" in turn contains another verb phrase (e.g. *reused for irrigating green space*) in the following sentence.

   "I want to see more rainwater reused for irrigating green spaces"

3 *Subject, object, and their relations* are extracted by the Stanford NLP Open Information Extractor. The dependency graph indicates 'I' as the subject and

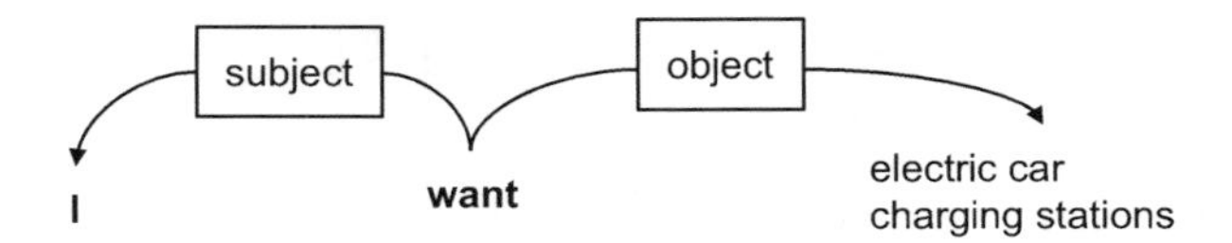

FIGURE 8.4 A Dependency Graph for "I want electric car charging stations".

'electric car charging stations' as the object, and the verb 'want' connects the subject and object, as shown in Figure 8.4.

4 *Named entities* are the names of things, detected by Stanford NLP Named Entity Recognizer (NER) in a text. Entities are labelled with entity types such as Person (e.g. Jacinda Ardern), Title (e.g. prime minister) or country (e.g. New Zealand).

Phase 3 - Thematic interpretation

Phase 3 provides a thematic interpretation of the augmented text data by attempting to differentiate between expressions of shared 'interests' compared with 'fixed positions' linked to thematic topics pertinent to CE.

Differentiating between 'interests' and 'fixed positions'

To differentiate between participants' 'interests' compared to 'fixed positions', all the sentences were examined for phrases beginning with the phrases "I want", "I like/love" or "I believe/think". The verb 'want' was interpreted as stating a 'fixed position' compared with the verbs 'like/love/believe/think' labelled as expressing an 'interest. The work by Nabatchi (loc cit) deemed it important to differentiate between these two standpoints because 'fixed positions' commonly lead to adversarial environments, whereas expressions of 'interest' foster collaborative work. The following example sentences from Christchurch 'Share an Idea' data illustrated 'fixed positions'.

> "I want recycling bins and more trees."
> "I want to see more rainwater reused for irrigating green spaces."

Whereas shared 'interests' expressed using the phrases *"I + like/love" or "I + believe/think" plus an object clause* are shown below.

> "I would like to use some of the bricks and material from buildings lost in the quake in new buildings."
> "I would love to see the inner city car free with a lot of cycle ways, bus lanes and pedestrian only areas."
> "I think that wind and solar generators should be installed on ALL CBD buildings."

"I believe that more green alternatives should be utilised to improve essential services e.g. power and sewerage."

Exploring thematic topics

Next, sentences were examined in relation to a predetermined glossary of terms used to characterise functions of urban infrastructures. The glossary of terms were generated using 'word embedding' techniques to represent words as vectors to calculate surrounding words with the similar meaning or similar vector score. For each two seed terms fed into the 'word embedding' model, ten similar or related words were output. For example, [train and rail] were combined to search for the similar words and produce a collection of representative terms of public transport, comprising express, tram, freight, railway and intercity. Three popular word-embedding models were trialled, namely: Wikipedia Word2Vec (Fares et al. 2017), Google News and Twitter and out of these, Wikipedia Word2Vec was found to produce the best results.

Phase 4 – Communication and visualisation

Phase 4 visualises the results from Phase 3 using a combination of chord charts, word trees, design personas and design briefs that together can compose a common narrative using data storytelling techniques. The first two techniques are illustrated in the following case study for Christchurch 'Share an Idea'. Examples of design personas and design briefs for use in collaborative urban design are presented in Dyer et al. (2019).

Results

Identify shared 'interests' or 'fixed positions'

Having extracted and 'cleaned' the data set, the linguistically augmented data set was used to identify shared 'interests' or 'fixed positions' about topics relevant to rebuilding of Christchurch that included waste, water, transportation and energy. Individual contributions were extracted using an automated glossary of terms together with personal pronouns and verbs that expressed personal "interests" in the form of "I love/like …" and "I think/believe …" compared with 'fixed positions' in terms of "I want …". A selected set of contributions are shown in Figures 8.5 and 8.6. The information provides a useful insight into topics of interest for the Christchurch community as well as proposed solutions. In terms of waste minimisation, Figure 8.5 reveals a wide interest in alternative ways of decentralising treatment of sewage using composting methods or collecting rainwater for household consumption. Likewise, suggestions were made about reusing demolition materials for rebuilding of the central city.

"I would **like** to **use** some of the **bricks** and material from buildings lost in the quake in new buildings so we can still have our heritage and remember."

"**I want** More green space integrated with **stormwater** treatment systems (the Avon River currently receives vast amounts of untreated stormwater filled with various contaminants)."

"**I want to see composting** toilets being the norm in the whole of Christchurch, they need to be subsidised by the CCC so that everyone can afford to put one into their home or business. This reduces the need for a complex and very costly sewerage system."

"**I want** Decentralisation of the **sewage** system. Alternative disposal methods investigated and implemented. Alternate energy systems devised

"**I want** each **house** to have a minimum of a 1,000 litre drinking **water** tank, a 30,000 litre tank for garden use, shared between 2 or 3 houses."

FIGURE 8.5 Christchurch 'Share an Idea' selected public contributions on topic of waste management.

"**I want cycleway**s along the river, to encourage cyclists and not cars."

"**I want** free **bike** usage for central city transport"

"I would **love** to **see** the inner **city** car free with a lot of cycle ways, bus lanes and pedestrian only areas."

"**I want** integrated **bus**, light rail, rail tram and bicycle transport."

"**I want** electric **buses**, within four avenues to reduce pollution and noise"

"**I want** little mini electric **buses** for transport around around the inner city and remove all cars"

"**I want** cycle lanes, **pedestrian**, car parking, public transport, cycle parking."

"**I want pedestrian** friendly streets, priority over cars when crossing streets, like in new york city"

"**I want** public **transport** that is more reliable than a bus light rail monorail subway etc free car parks in outlying suburbs to allow commuters and workers to drive to these points for easy and fast access to the centre city"

"**I want** a **transport** network that makes it easy to use public transport to get into the city.with children/pushchair."

FIGURE 8.6 Christchurch 'Share an Idea' selected public contributions on modes of transportation.

Similarly, Figure 8.5 shows a public appetite for more public transport, cycling, pedestrianisation and provision of electric vehicles together with sharing of vehicles. However, listing individual contributions in this way provides only a fragmented view of the community engagement, well removed from the conversation proposed by Christchurch City Council when launching 'Share an Idea'. To examine how a two-way conversation could be facilitated, the study explored the use of 'Word Trees' and 'Chord Charts' to visualise future community conversations.

Christchurch 'word tree' for CE

To explore community conversations from Christchurch 'Share an Idea' about future management of waste, water and transportation for the Central City Centre, 'Word Trees' were constructed using Google Chart. The diagrams were plotted using sentences comprising expressions of 'interests' in the form of "I like/love ..." or "I think/believe ..." compared with 'fixed positions' conveyed as "I want ...". The resulting Word Trees are illustrated in Figures 8.7 and 8.8 for waste management and transportation, respectively. Even though the public consultation was centred around the broad topic of rebuilding the city centre after major earthquakes, the topics of interest were very pertinent to future sustainable development of the city. In the case of waste management, the Word Tree in Figure 8.7 indicates a strong level of feeling towards 'wanting' change with an emphasis on sewage and composting. In comparison, Figure 8.8 indicates an interest in replacing cars with more cycling, walking and public transport that is ideally electric driven. Furthermore, there were comments on more efficient use of space, especially for car parking. All of these suggestions for waste minimisation and alternative modes of transport align with opportunities to increase sustainable development.

Christchurch 'chord chart'

The second approach used to interpret key messages from Christchurch 'Share an Idea' involved constructing Chord Charts. The images were plotted using D3.js open source library to illustrate inter-relationships between different aspects of urban infrastructures that impact on sustainable development. The terms 'Soft and Hard Urban Infrastructures' were introduced by Dyer et al. (2019) to represent the physical utilities, buildings and spaces as 'Hard Infrastructure' compared with administrative, social and personal characterised categorised as 'Soft Infrastructure'. The approach was adopted because an initial examination of 'Share an Idea' data set showed a public awareness of connections or disconnections between 'Soft and Hard Urban Infrastructures' that affected livelihood and liveability, such as night shift workers needing to travel home or single parents with children needing easy access to recreation facilities and schooling.

The resulting Chord Charts from Christchurch 'Share an Idea' are displayed in Figure 8.9. The main Chord Chart on the left-hand side illustrates the multiplicity of views expressed connecting different aspects of 'Soft and Hard Urban Infrastructures'. The width of the chord indicates the relative proportion of contributions received for that particular combination of 'Soft and Hard Urban Infrastructure'. For example, there is a noticeably wide chord representing participant views about connections between People as 'Soft Infrastructure' and transport/buildings as 'Hard Infrastructure'. The online application allows the individual contributions to be viewed by clicking on the relevant chord. For example, the Chord connecting People and Transport captured views and preferences about different modes of

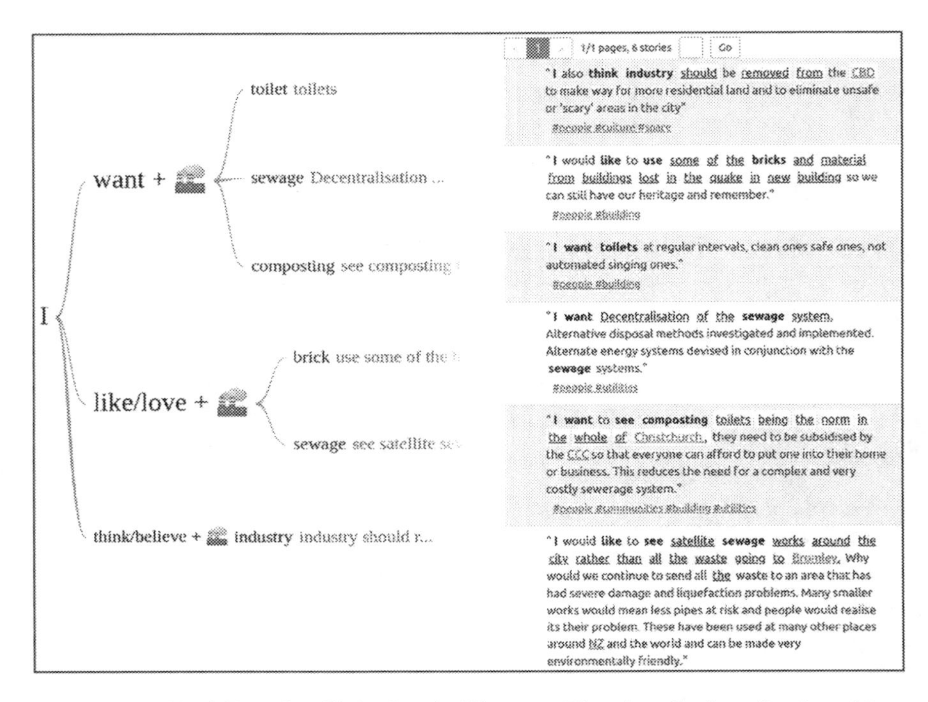

FIGURE 8.7 Word Tree for Christchurch 'Share an Idea that displays fixed positions expressed as "I want …" compared with shared interests conveyed as "I like/love…" or "I think/believe …" about waste minimisation and reuse.

transportation as documented in Figure 8.8. Ultimately, both graphical techniques allow public contributions to be processed and displayed in real time as part of a dynamic deliberative process.

Discussion

Framework for using AI NLP tools to enhance public consultation

The study has demonstrated some of the capabilities of AI NLP tools to enhance the interpretation of large format text data from public consultation but also revealed technological and political limitations that should be discussed in more detail.

On a positive note, the AI NLP toolkit called Urban Narrative offers a quick and easy means of reaggregating individual POSs and clauses from original text, whether collected online or in person, into a coherent set of statements based on preferred combinations of nouns, verbs and adverbs that reflect Shared Interests, such as personal noun (*I or we*), common noun (*waste or transportation*), proper nouns (*Christchurch*), verbs (*love, hate*) and adverbs (*must* have). The approach enables a common narrative to develop around Shared Interests or Fixed Positions,

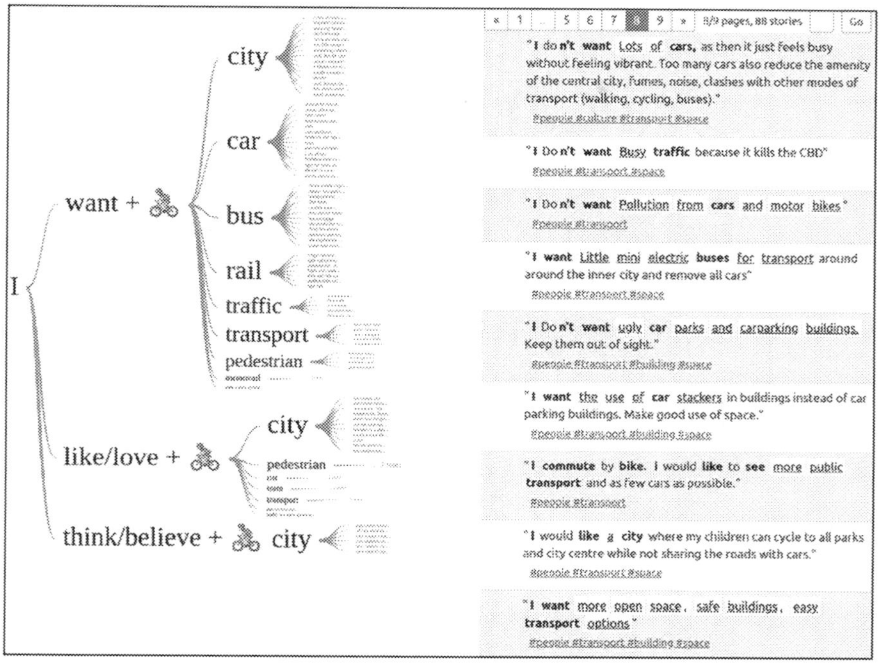

FIGURE 8.8 Word Tree for Christchurch 'Share an Idea' displaying fixed positions expressed as "I want …" compared with shared interests conveyed as "I like/love…" or "I think/believe …" about alternative modes of transportation in the central city.

which in the Case Study concerns the rebuilding of Christchurch City Centre. From previous studies by the author (Dyer et al. 2017) and others (Nabatchi 2012, Fung 2006), the creation of a common dialogue about Shared Interests was found to be an essential ingredient for successful public consultation that helps define a common understanding of the issue at hand rather than jumping to early solutions that might not be relevant but instead cause conflict and misunderstanding.

Likewise, the case study revealed the importance of framing the issue at hand for Public Consultation into a suitable question that encourages discussion about Shared Interests instead of asking for preferred solutions that lead to Fixed Positions. In this instance, the phrase 'Share an Idea' used in the Case Study encourages people to propose their personal solution rather than share ideas or thoughts about abstract ideas that could otherwise be known as Shared Values. Instead, the Case Study unwittingly encouraged the public to take up 'fixed positions' on topics using the phrase "I want …." compared with shared 'interested' expressed using phrases "I love/like …" or "I think/believe …". At the same time, it was intriguing to observe that phrases "I want …" commonly related to objects, whereas the phrases "I think/believe …" or "I love/like …" often related to less tangible issues such as

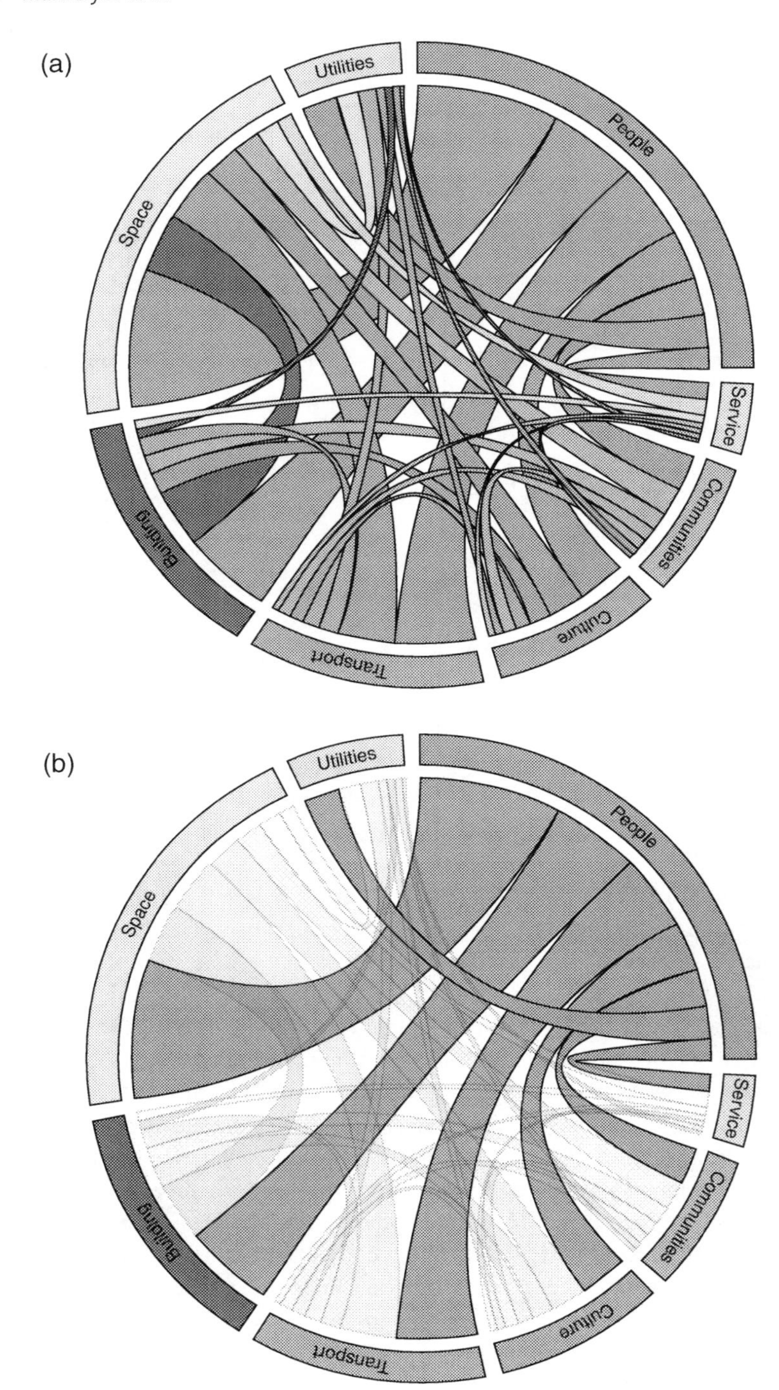

FIGURE 8.9 Chord Chart displaying perceived connections between 'Soft and Hard Urban Infrastructures' expressed in public contributions for Christchurch 'Share an Idea'.

environment or aesthetics. The result indicates a need for careful design of public consultation to encourage greater expressing of Shared Interests rather than Fixed Positions. With hindsight, it would have been more valuable to seek the public's cultural values and beliefs first and then encourage ideas or solutions about desired qualities for sustainable reconstruction. This would provide designers (e.g. architects, urban planners, engineers or product designers) with more accurate information and knowledge about 'users'' needs to enable them to ideate people-centred solutions.

As an inspiration for using Cultural Values in Urban Design, New Zealand's Maori community of urban designers have recently embraced indigenous cultural values to establish a set of core Urban Design Principles as shown in Table 8.1. Entitled *Te Aranga Seven Māori Design Principles* (Awatere et al. 2010), the design principles are embedded in Maori Cultural Values and implicitly lead to discussions about society's core values before identifying associated design solutions that turn Values into Qualities. By that, I mean for example that emphasis on the principle Kaitiakitanga would imply that protection of the environment and ecology is a major Value that needs to be addressed in any design solutions such as promoting ecological diversity in suitably designed greenspaces or avoiding pollution of groundwater. Likewise, the principle Manakitanga would imply that urban design solutions would embrace hospitality for residents and visitors to those neighbourhoods such as providing meeting places. The issues might appear to be nebulous or abstract principles but earlier studies by the author have shown that core values can act as key motivators for significant change, such as the

TABLE 8.1 Te Aranga Māori Urban Design Principles (after Awatere et al. 2010)

Rangatiratanga (Ownership) The right to exercise authority and self-determination within one's own Iwi / Hapū (Tribe/Sub-Tribe).	**Kaitiakitanga (Guardianship)** Managing and conserving the environment as part of a reciprocal relationship, based on the Māori world view that we as humans are part of the natural world
Manakitanga (Hospitality) The ethic of holistic hospitality where by mana whenua have inherited obligations to be the best hosts they can be	**Wairuatang (Spirituality)** The immutable spiritual connection between people and their environments
Kotahitanga (Unity) Unity, cohesion, and collaboration	**Whanaungatanga (Relationships)** A relationship through shared experiences and working together which provides people with a sense of belonging
Mātauranga (Knowledge) Māori / mana whenua knowledge and understanding	

transition of two Scandinavian Cities of Vaxjo and Sonderborg towards becoming low-carbon societies (Dyer and Ögmundardóttir 2018). Likewise for the Case Study, a focus on Shared Values would have shown that participants prioritised the environment and outdoor exercise as principles first before deciding on creation of large green spaces for walking, cycling and exercise as design solutions. Hence, AI NLP can be seen as a vehicle for promoting a more nuanced discussion about Shared Values before selecting design solutions as part of a Collaborative Design Framework.

Furthermore, the Case Study shows the importance of establishing guidelines for shared decision making from the outset. Otherwise, public participants can become disillusioned when their ideas and recommendations are not given proper consideration when decisions are taken by a remote governing body. This might seem an obvious point to raise but it is fundamental to the success and authenticity of public engagement. The reluctance to shared decision making also loses the opportunity to share knowledge and wisdom beyond the narrow confines of officialdom and professional bodies that can be prone to groupthink as witnessed by some disastrous mistakes in urban design since mid-twentieth century (Jacobs 1961). Lessons learnt from Christchurch 'Share an Idea' show that the online public consultation was readily supported by the community, albeit under emergency conditions following the devastating earthquakes. The grassroots process proposed several important design preferences for the rebuild such as limiting the building height in the Central City to six stories. However as explained by Brand and Nicholson (2016), many of the grassroots ideas stemming from the grassroots initiative fell afoul of a top-down central government bureaucracy. In this case, CERA took complete control for the Recovery Plan through a top down exercise conducted behind closed doors (Gjerde 2017). The bureaucratic process attracted considerable criticism with subsequent research revealing that CERA was both incapable and unwilling to engage effectively with the public (Simons 2016). Not surprisingly, public perceptions of the way the recovery was managed were perceived as being very poor, with up to 80% of the people living in Christchurch holding a negative view (Simons 2016). Reasons for this are many and are likely to reference a person's experiences of dealing with their own circumstances. But overall, Christchurch 'Share an Idea' appears to have been a lost opportunity to facilitate a genuine collaborative process for the rebuild that could readily embrace many of the underlying concepts of circular economy (CE) embedded in the public consultation as documented in Figures 8.7 and 8.8.

In marked contrast, the strategies adopted by the cities of Växjö and Sønderborg emphasised the importance of ongoing public consultation and engagement to reinforce community narratives based on core values that underpinned one of the pillars of sustainable development. Based on these lessons learnt, in each case the convergence of Public Consultation with Participatory Design to motivate a change in public mindset needs a transparent sharing of power between 'top and bottom-up' processes to facilitate ongoing two-way deliberative discussion supported by willingness to trial and modify novel solutions. This is where AI NLP-driven real-time public consultation provides a means to create ongoing community narratives based on shared 'interests' to aid design of public policy and design of new services, products and infrastructure.

Lastly, there is a need to recruit representative stakeholders to embrace diversity and avoid bias. Of course, the level and diversity of engagement for public consultation exercises can vary considerably ranging from state-based participants (e.g., expert administrators and elected representatives) to mini-publics (e.g., professional or lay stakeholders, or randomly selected, self-selected or recruited individuals) to diffuse members of the public (Fung 2006). The use of NLP implies an online recruitment which can influence the type of participants. In the case of Christchurch 'Share an Idea', online engagement was supplemented by in person engagement (via town hall meetings, interviews and written submissions) to broaden the range of participants. Attracting engagement from a broad range of participants is crucial to foster transformational change at scale. It becomes even more important when attempting to construct design personas that represent different socio-demographic groupings. Partly for that reason, it would be valuable where possible to connect this type of public consultation exercise with census data subject to ethical and privacy approval.

In summary, a framework for successful implementation of AI powered Public Consultation require the following key elements based on the results of the Case Study, i.e.

- Framing of question(s) to encourage discussion about Shared Interests and Common Values instead of asking for preferred solutions that can lead to Fixed Positions.
- Ensuring diverse Public Engagement with representative stakeholders to avoid bias, particularly when relying more on online engagement vis-à-vis in person consultation.
- Agreeing from the outset the degree to which Shared Decision making is permitted to avoid future disillusionment
- Creating a coherent dialogue by selecting a preferred set of nouns, verbs and adverbs that can reaggregate POS into a coherent discussion about Shared Interests.

AI-powered public participation as a vehicle for large-scale participatory design

Viewing the study from another perspective, the information gathered on 'Shared Interests' can be seen as an unique opportunity for the public to be involved in large scale participatory design processes where the public as 'users' are considered a full design partner. Depending on the extent of sharing decision making, this approach has the potential to empower the 'users' to be active participants in an innovative design process. Hence the designer would become a facilitator or what Ehn (2008) describes as a responsive designer, one who alternates the leadership roles in a project depending on whose skills are most relevant whilst keeping all participants involved.

Intriguingly, Dewey (1981), as twentieth-century thought leader for public participation, implicitly recognised the overlap between public participation mechanisms and participatory design by once remarking that "the man who wears the shoe, not the shoemaker, knows best where it pinches". Consequently, there is a great deal of potential commonality between Public Participation and Participatory Design, especially where the two aim to empower the 'user'. This provides the theoretical foundations to explore ways of integrating both processes to facilitate a paradigm shift towards future sustainable development.

Furthermore, it is worth noting that participatory design is founded on two complementary values, the first being the right to participate in design activities and the second a means of bringing 'tacit' or non-discursive knowledge of users into design thinking. Hence, without adequate sharing of authority the public are potentially excluded from the design process and all the knowledge and insight that they can bring. To be successful Participatory Design requires sharing of authority and recognition that a design process is iterative. In practice, this means devising feedback systems to facilitate a two-way dialogue that is valid, respectful and representative of the community whilst creating a less static and more dynamic environment to trial prototypes before selecting more permanent solutions.

At the same time, it is important to recognise that design thinking is an iterative process that starts with understanding the needs of end users (Dyer, Corsini, and Certomà 2017). In participatory design, this is achieved by involving the user directly within the design team as a design partner, where the design partner identifies their needs in a design forum/workshop. The challenge for large-scale public initiatives is upscaling engagement beyond traditional numbers of participants from 10s or 100s to 10,000s plus. Taking inspiration from the success experienced by Växjö and Sønderborg, a digital platform(s) has been developed using NLP toolkits to upscale large-format public participation to identify public 'interests' and 'fixed positions' as a part of a people-centred design process for new products, services and infrastructure to aid sustainable development.

Conclusion

Inspired by Växjö and Sønderborg use of community narratives to transition towards a fossil fuel free cities, this study has demonstrated the potential use of NLP toolkits to upscale public consultation by analysing large format public text data to identify participants individual Shared Interests and Fixed Positions using graphical storytelling tools in the form of Word Trees and Charts. The result is a digital platform called Urban Narrative that can facilitate a community-wide discussion about a broad spectrum of complex issues in real-time. Beyond the scope of this study are other opportunities to link the data analytics to socio-demographic information about participants to create design personas that can aid participatory design processes but with the clear provision of needing ethical guidelines to protect personal privacy.

However, for the digital platform to be effective, there are a number of issues that need addressing. The first is sharing of authority for decision making. Governing bodies are often reluctant to share authority for a host of different reasons. Likewise, it is crucial to focus on the public's 'interests' in the form of Shared Interest and Beliefs rather than Fixed Positions. This means designing public consultation processes to guide participants towards firstly identifying interests in terms of underlying values by concentrating on the 'why' not 'what'.

Finally, the study highlighted the potential overlap between public participation processes and participatory design that can be integrated to engage in large-scale public discourse and re-evaluation to rethinking services and built environment amongst other aspects of sustainable development. In this scenario, the identification of participants' interests (values) and even positions (solutions) via NLP analysis provides the core insights about 'users' needs to aid design thinking to re-imagining future tangible and intangible product chains.

References

Awatere, S., Rolleston, S., & Pauling, C. (2010). Developing maori urban design principles. In K. Stuart & M. Thompson-Fawcett (Eds.), *Tāone tupu ora: Indigenous Knowledge and Sustainable Urban Design* (pp. 17–23). Wellington, New Zealand: Steele Roberts Aotearoa.

Brand, D., & Nicholson, H. (2016). Public space and recovery: Learning from post-earthquake Christchurch. *Journal of Urban Design*, *21*(2), 159–176. doi: 10.1080/13574809.2015.1133231

Christchurch City Council. (2011). *Christchurch Central Recovery Plan*. Available at: https://ccc.govt.nz/assets/Documents/The-Council/Plans-Strategies-Policies-Bylaws/Plans/central-city/CentralCityPlanTechnicalAppendicesA-D.pdf (Accessed November 1, 2023).

Dalton, T., Draper, M., Wiseman, J., & Weeks, W. (2020). *Making Social Policy in Australia: An Introduction*. New York: Routledge.

Dewey, J. (1981). *The Later Works of John Dewey, Volume 2, 1925—1953: 1925-1927, Essays, Reviews, Miscellany, and the Public and Its Problems* (J. A. Boydston, Ed.). Carbondale: Southern Illinois University Press.

Dyer, M., Corsini, F., & Certomà, C. (2017). Making urban design a public participatory goal: Toward evidence-based urbanism. *Proceedings of the Institution of Civil Engineers - Urban Design and Planning*, *170*(4), 173–186. doi: 10.1680/jurdp.16.00038

Dyer, M., Dyer, R., Weng, M.-H., Wu, S., Grey, T., Gleeson, R., & Ferrari, T. G. (2019). Framework for soft and hard city infrastructures. *Proceedings of the Institution of Civil Engineers - Urban Design and Planning*, *172*(6), 219–227. doi: 10.1680/jurdp.19.00021

Dyer, M., Gleeson, D., Ögmundadottir, H., Ballantyne, A. G., & Bolving, K. (2017). Awareness, communication and visualisation. In M. E. Goodsite & S. Juhola (Eds.), *Green Defense Technology* (pp. 269–286). Springer Netherlands. doi: 10.1007/978-94-017-7600-4_13

Dyer, M., & Ögmundardóttir, H. (2018). Transition of Växjö and Sønderborg towards becoming fossil-fuel-free communities. *Proceedings of the Institution of Civil Engineers - Energy*, *171*(1), 3–11. doi: 10.1680/jener.17.00004

Ehn, P. (2008). DOC and the power of things and representatives. *Proceedings of the 26th Annual ACM International Conference on Design of Communication*, 31–32. doi: 10.1145/1456536.1456543.

Eisenstein, J. (2019). *Introduction to Natural Language Processing*. Cambridge & London: MIT Press.

Enli, G. (2017). Twitter as arena for the authentic outsider: Exploring the social media campaigns of Trump and Clinton in the 2016 US presidential election. *European Journal of Communication*, *32*(1), 50–61. doi: 10.1177/0267323116682802

Fares, M., Kutuzov, A., Oepen, S., & Velldal, E. (2017). Word vectors, reuse, and replicability: Towards a community repository of large-text resources. *Proceedings of the 21st Nordic Conference on Computational Linguistics*, 271–276. Available at: https://aclanthology.org/W17-0237 (Accessed November 1, 2023).

Fischer, F. (2003). *Reframing Public Policy: Discursive Politics and Deliberative Practices*. Oxford University Press. doi: 10.1093/019924264X.001.0001

Fung, A. (2003). Survey article: Recipes for public spheres: Eight institutional design choices and their consequences. *Journal of Political Philosophy*, *11*(3), 338–367. doi: 10.1111/1467-9760.00181

Fung, A. (2006). Varieties of participation in complex governance. *Public Administration Review*, *66*(s1), 66–75. doi: 10.1111/j.1540-6210.2006.00667.x

Gildea, D., & Jurafsky, D. (2002). Automatic labeling of semantic roles. *Computational Linguistics*, *28*(3), 245–288. doi: 10.1162/089120102760275983

Gjerde, M. (2017). Building back better: Learning from the christchurch rebuild. *Urban Transitions Conference, Shanghai, September 2016*, *198*, 530–540. doi: 10.1016/j.proeng.2017.07.108

Gleeson, D., & Dyer, M. (2017). Manifesto for collaborative urbanism. In C. Certomà, M. Dyer, L. Pocatilu, & F. Rizzi (Eds.), *Citizen Empowerment and Innovation in the Data-Rich City* (pp. 3–18). Springer International Publishing. doi: 10.1007/978-3-319-47904-0_1

Hinton, G. E., Osindero, S., & Teh, Y.-W. (2006). A fast learning algorithm for deep belief nets. *Neural Computation*, *18*(7), 1527–1554. doi: 10.1162/neco.2006.18.7.1527

Jacobs, J. (1961). *The Death and Life of Great American Cities*. Random House Publishers. ISBN 0-679-74195-X

Kalchbrenner, N., & Blunsom, P. (2013). Recurrent continuous translation models. *Proceedings of the 2013 Conference on Empirical Methods in Natural Language Processing*, 1700–1709. Available at: https://aclanthology.org/D13-1176.pdf (Accessed November 1, 2023).

Manning, C. D. (2015). Last words: Computational linguistics and deep learning. *Computational Linguistics*, *41*(4), 701–707. doi: 10.1162/COLI_a_00239

Manning, C., Surdeanu, M., Bauer, J., Finkel, J., Bethard, S., & McClosky, D. (2014). The stanford CoreNLP natural language processing toolkit. *Proceedings of 52nd Annual Meeting of the Association for Computational Linguistics: System Demonstrations*, 55–60. doi: 10.3115/v1/P14-5010

Nabatchi, T. (2012). Putting the "public" back in public values research: Designing participation to identify and respond to values. *Public Administration Review*, *72*(5), 699–708. doi: 10.1111/j.1540-6210.2012.02544.x

Nielsen. (2018). *Quality of Life Survey 2018: Topline Report. A Report Prepared on Behalf of Auckland Council, Wellington City Council, Christchurch City Council, and*

Dunedin City Council. Nielsen, Wellington, New Zealand. Available at: https://www.qualityoflifeproject.govt.nz/wp-content/uploads/2022/10/Quality-of-Life-Technical-Report-2018.pdf (Accessed November 1, 2023).

Sakaki, T., Okazaki, M., & Matsuo, Y. (2010). Earthquake shakes Twitter users: Real-time event detection by social sensors. *Proceedings of the 19th International Conference on World Wide Web*, 851–860. doi: 10.1145/1772690.1772777

Sam, M. P. (2003). What's the big idea? Reading the rhetoric of a national sport policy process. *Sociology of Sport Journal*, *20*(3), 189–213. doi: 10.1123/ssj.20.3.189

Simons, G. (2016). Projecting failure as success: Residents' perspectives of the Christchurch earthquakes recovery. *Cogent Social Sciences*, *2*(1), 1126169. doi: 10.1080/23311886.2015.1126169

Socher, R., Bauer, J., Manning, C. D., & Ng, A. Y. (2013). Parsing with compositional vector grammars. *Proceedings of the 51st Annual Meeting of the Association for Computational Linguistics (Volume 1: Long Papers)*, 455–465. Available at: https://aclanthology.org/P13-1045 (Accessed November 1, 2023).

Socher, R., Perelygin, A., Wu, J., Chuang, J., Manning, C. D., Ng, A., & Potts, C. (2013). Recursive deep models for semantic compositionality over a sentiment treebank. *Proceedings of the 2013 Conference on Empirical Methods in Natural Language Processing*, 1631–1642. Available at: https://aclanthology.org/D13-1170 (Accessed November 1, 2023).

Tambouris, E., Macintosh, A., Dalakiouridou, E., Smith, S., Panopoulou, E., Tarabanis, K., & Millard, J. (2013). eParticipation in Europe: Current state and practical recommendations. In J. R. Gil-Garcia (Ed.), *E-Government Success around the World: Cases, Empirical Studies, and Practical Recommendations* (pp. 341–357). IGI Global. doi: 10.4018/978-1-4666-4173-0.ch017

Wang, H., Can, D., Kazemzadeh, A., Bar, F., & Narayanan, S. (2012). A system for real-time Twitter sentiment analysis of 2012 U.S. presidential election cycle. *Proceedings of the ACL 2012 System Demonstrations*, 115–120. Available at: https://aclanthology.org/P12-3020 (Accessed November 1, 2023).

9

DATA (UN)SUSTAINABILITY

Navigating utopian resistance while tracing emancipatory datafication strategies

Igor Calzada

We have a big (tech) problem

In an era where the digital landscape evolves at warp speed, the reliance of digital citizens on big tech has raised a complex issue that demands our undivided attention (Calzada 2022). These disruptive technologies, as a result of creative destruction processes (Schumpeter 1942), have brought immeasurable benefits to billions of people, including improved health, employment, and wellbeing (Burr and Floridi 2020). Especially during times of crisis, disruptive technologies have played an increasingly critical role in human and societal survival, particularly within the contemporary political economy that encompasses various relational and institutional aspects of capitalism (Polanyi 1944). Disruptive technologies have played a pivotal role, for example, in global conflicts and natural disasters, including extreme weather events that have resulted in the displacement of large numbers of people. Furthermore, disruptive technologies in the pandemic era have also transformed communities and ways of living and working resulting in a new pattern known as 'pandemic citizenship' (Bignami et al. 2022; Calzada 2022). However, these disruptions can also lead to unforeseen destructive consequences. The harms of dominant and data-opolytic technology platforms are manifold (Stucke 2022). They include the exploitation of data, impacts on the mental health and safety of minors, the proliferation of misinformation, and adverse effects on political institutions and behaviour. Big tech, particularly social media companies, has thus become the subject of public scrutiny and criticism. Hence, both internal company initiatives and external bipartisan attempts to address these issues have met with limited success (Spelliscy et al. 2023; Srivastava 2021).

Consequently, during and after the pandemic, data flows, transfers, migrations, and algorithmic disruptions have become commonplace, impacting citizens' digital

DOI: 10.4324/9781003441311-11

rights and undermining their data privacy (van Dijck 2018). This mainstream data extractivism, extending across deep, biometric, and postpandemic borders, places digital citizens, in particular, at a greater risk of data privacy breaches by revealing a totalitarian order that has become dominant in the global digital landscape (Arendt 1966). Data extractivism refers to the practice of collecting, analysing, and commodifying large amounts of personal data from digital citizens without their explicit consent or control, often for commercial or political purposes (Sadowski 2019). This practice involves the use of digital technologies, such as social media platforms, to gather personal data, including online behaviour, preferences, and demographic information, and convert it into an asset for companies and governments (O'Shea 2021). Consequently, data extractivism poses a challenge to the ethical and democratic governance of datafied societies, emphasizing the imperative to protect the digital rights of digital citizens (Zuboff 2019).

Moreover, Stucke poses an insightful question: Why have Google, Apple, Facebook, Amazon, and Microsoft (GAFAM) successfully dominated multiple markets for years and seem poised to continue their domination over the next decade? These data-opolies have controlled the digital economy, and 'the price we pay includes our privacy, attention, and autonomy' (Stucke 2022: 1). According to Stucke, four well-accepted factors explain this data-opolistic dominance trend that impacts digital citizens' data (un)sustainability (European Commission 2020): (i) economies of scale, (ii) network effects, (iii) attention, and (iv) the four Vs of personal data, which stand for volume, variety, velocity in processing, and value. Once a data-opoly has achieved such economies of scale and established a network effect, it becomes increasingly difficult for new entrants to attract a substantial user base. Network effects occur when a product or service's value increases as others use it. There are five network effects in the digital platform economy: (i) the direct network effect, (ii) the indirect network effect, (iii) spillover effects, (iv) the learning-by-doing effect, and (v) the scope of data network effect. Utopian resistance movements should clearly address this global challenge beyond the regulatory frameworks that multiply continually. Data-opolies can thus create (or harness) these network effects for their advantage and to lock us in.

Against this backdrop, scholars such as Bucher (2012), Forestal (2020), and Taplin (2017) argue that data-opolies or big tech platforms, particularly Google and Facebook, employ algorithms and opaque content moderation policies to obscure the prioritization and promotion of content on their platforms. Critics assert that this practice has the potential to amplify misinformation and create echo chambers. They further argue that such an approach conceals the true nature of presented content, eroding public trust and exacerbating the polarization of public opinion. On the other hand, proponents of these platforms contend that they prioritize free speech and user autonomy, while also recognizing the need to address concerns such as misinformation and harmful content. Veliz (2020) suggests that enhancing transparency and accountability in the algorithms and content moderation policies of these platforms could help alleviate these privacy concerns. However, Gorwa

(2019) posits that regulating these platforms is a multifaceted and intricate issue, demanding a nuanced approach.

Considering the arguments and counter-arguments in this ongoing debate, an alternative and widespread reaction emerges from crypto-libertarian or pseudo-anarchist perspectives. This reaction has led to the development of an emerging body of literature on decentralized systems in peer-to-peer interactions, encompassing (i) blockchain (Calzada 2023a; Hall, 2023), (ii) DAOs (Dupont 2017; Hubbard 2023), and (iii) data co-operatives (Bauwens et al. 2019; Bühler et al. 2023b; Buterin 2022; Calzada 2021a; Mathew 2016; Monsees 2019; Rennie et al. 2022). This alternative viewpoint suggests an unexplored research trajectory that could interweave blockchain, DAOs, and digital citizens, potentially fuelling a post-identitarian mobility pattern (Inwood and Zappavigna 2021; Isin and Ruppert 2015). It is essential to acknowledge that this alternative stance, currently advocating for blockchain and DAOs, finds its origins in *The Crypto Anarchist Manifesto* launched in 1988 in Silicon Valley by Timothy May (Nabben 2022).

Consequently, in the face of the formidable challenges posed by big tech's unchecked data practices, a beacon of hope emerges on the digital horizon in the form of Web3 technologies. Web3, driven by the principles of decentralization, transparency, and data sovereignty, offers a potent antidote to the unsustainable data practices that have come to define the digital age. At its core, Web3 represents a fundamental shift in how we conceive and manage data, fostering a sustainable vision of the digital future. In the world of Web3, blockchain technology, underpinning cryptocurrencies like Bitcoin and Ethereum, takes centre stage as a decentralized ledger that ensures data integrity and immutability (Hughes et al. 2019). This digital foundational economic model empowers individuals, allowing them to reclaim their digital identities from the grasp of monolithic tech giants (Morozov 2022; Morozov and Cancela 2023).

However, Web3s true revolutionary potential extends beyond blockchain (Viano et al. 2023; Zook 2023). It gives birth to DAOs, autonomous entities governed by code and consensus, where decisions are made collectively by stakeholders rather than dictated by corporate hierarchies (Stanford DAO Workshops 2022 and 2023; WEF 2023). This shift towards decentralized governance challenges the very core of big tech's dominance, offering a more equitable and democratic approach to data management (Toscano 2021). In this digital utopia, data cooperatives emerge as key players, enabling individuals to collectively manage and profit from their data (Bühler et al. 2023a; Calzada 2021a). Digital citizens are no longer mere consumers; they become active participants, shaping the rules and benefits of the data ecosystem of which they are a part (Bignami et al. 2022; Calzada 2018; Farrell et al. 2023).

What makes Web3 truly utopian is its inherent resistance factor. It is a grassroots movement that resists the centralization of data power and envisions a world where digital citizens regain control over their online lives. Web3 pioneers a sustainable ethos that champions data privacy, environmental responsibility, and economic fairness. By aligning with Web3, we embark on a journey towards a digital future where the unsustainable practices of big tech are replaced by a more equitable, sustainable, and democratic data ecosystem. The promise of Web3 lies not only in its technological innovations but also in the empowerment of digital citizens to shape a future where data serves the collective good, where sustainability, privacy, and resilience are paramount (Burrell and Fourcade 2021). In this way, Web3 charts a path towards a digital utopia, where people reclaim their data destiny and forge a more sustainable and just digital world.

Hence, this chapter encapsulates the urgency of our times—a call to reckon with the collision of utopian dreams and the harsh reality of unchecked technological power. In the pages that follow, we embark on a journey that seeks not only to illuminate the challenges but also to elucidate potential pathways for joint action research and policy. We aim to inspire readers to join the conversation and contribute to the urgent task of shaping the digital world we want and need.

Utopian resistance against data extractivism

With this literature review, we embark on an odyssey to decode the intricate script of data's destiny. This is an exploration that delves beyond the surface, uncovering not only the quandaries but also the pathways to redemption. As we traverse the precipice of possibility, we are tasked with steering the course towards a digital future that champions emancipation over subjugation, sustainability over recklessness—a future that stands as a testament to our ability to conquer the colossal (tech) problem that looms large. In an era marked by unprecedented technological advancements, our world has become tightly intertwined with the digital realm. The influence of technology giants and their data-driven ecosystems has infiltrated every aspect of our lives, reshaping societies and economies with unprecedented force. As we navigate this complex landscape, it becomes increasingly evident that the promises of a utopian digital future, once proclaimed by visionary thinkers like John Perry Barlow (1996), have encountered substantial turbulence on their journey to realization. The interplay between data, sustainability, democracy, and emancipatory strategies stands at the forefront of our contemporary discourse, demanding urgent attention and collective action. As such, John Perry Barlow's Declaration of the Independence of Cyberspace in 1996 projected a vision of a digital realm free from the constraints of physical borders, a utopia where information would flow freely, transcending traditional hierarchies and power structures. However, as the years unfolded, this vision collided with the reality of big tech's monopolistic practices, privacy breaches, and the rise of surveillance capitalism, all of which stand as stark reminders of the unfulfilled utopia (Stucke 2022).

In the shadows cast by the 'dark side' of technology, James Bridle's expose on the hidden infrastructure of the digital world unearths troubling realities that lie beneath the surface (2019). The tension between utopian aspirations and dystopian consequences invites us to reconsider the narrative surrounding our digital existence. Similarly, Ekaitz Cancela's work on 'Utopías Digitales' urges us to critically examine the potential and limitations of the digital sphere as a tool for societal transformation (2023). It prompts us to explore strategies that might enable us to harness the emancipatory potential of technology while addressing the pressing issues of data sustainability. Javier Echeverría's concept of 'Telépolis' further deepens this exploration (1994). The idea of a technologically enabled global village offers glimpses of connectivity and collaboration, but it also demands careful consideration of the environmental, social, and ethical implications. As we grapple with the all-encompassing reach of technology, we find ourselves in need of a new paradigm that aligns digital innovation with sustainability goals. Cal Newport's (2019) proposition of 'Digital Minimalism' suggests a recalibration of our relationship with technology to undermine hyperconnectivity, as advocated by Calzada and Cobo (2015) in their article 'Unplugging.' It encourages intentional and mindful use to counter the digital overload that threatens to engulf us (2019). To navigate the intricate web of issues surrounding data (un)sustainability, we must also address the economic dimensions. Lizzie O'Shea's concept of 'Future Histories' reminds us that the choices we make today in shaping our digital landscape will reverberate through history (2021). This realization calls for strategic foresight and a proactive approach to crafting policies and systems that withstand the test of time. Markku Lehdonvirta's exploration of 'Cloud Empires' delves into the power dynamics of data accumulation, shedding light on the concentration of authority within the hands of a few corporate behemoths. As we seek to foster a more sustainable digital ecosystem, it becomes imperative to rebalance these power structures and cultivate equitable data governance mechanisms.

In this digital odyssey, the writings of Hannah Arendt in 'The Origins of Totalitarianism' (1966) serve as a profound historical context for understanding the complex interplay between technology, power, and societal transformation. Arendt's exploration of the rise of totalitarian regimes in the twentieth century provides a cautionary tale that resonates with our contemporary digital age. Arendt's insights into the dangers of political and technological systems that undermine human agency are particularly relevant as we grapple with the consequences of data extractivism and surveillance capitalism in the digital realm, emphasizing the 'right to have digital rights' (Calzada 2021b). In this literature review, Arendt's perspective acts as a historical anchor, reminding us that the struggle for a sustainable digital future is not a new one (O'Shea 2021).

Emerging digital citizenship regimes

In an era marked by global connectivity and digital interdependence, traditional notions of citizenship have transcended physical borders (Cheney-Lippold 2016; De Filippi and Schingler 2023). The post-pandemic technopolitical democracies explored by Calzada illustrate a shift towards a digital-centric citizenship, where participation in digital spaces becomes intrinsic to one's sense of belonging and agency. This transformation compels us to explore the potential implications and opportunities for emancipatory strategies within these new digital citizenship regimes (Marquardt 2021).

The proliferation of digital technologies has brought forth a profound transformation in the landscape of citizenship, extending beyond traditional notions of civic engagement and political participation. In 'Emerging Digital Citizenship Regimes,' Calzada provides a comprehensive analysis of the evolving paradigms of citizenship within the digital age. Central to this exploration is the concept of 'datafication,' a multifaceted phenomenon that implicates both the potential for emancipatory empowerment and the inherent perils of data-driven governance. Calzada's work contextualizes the contemporary discourse on digital citizenship within the broader framework of sustainability, elucidating the intricate relationships between data practices, societal values, and ecological considerations in relation to the post-westphalian nation-state. Calzada's work sheds light on the interconnectedness between technological advancements, political structures, and digital rights between existing disparities in the Global North and Global South (Couldry and Mejias 2019). 'Emerging Digital Citizenship regimes' identifies five ideal types of digital citizenship: pandemic, algorithmic, liquid, metropolitan, and stateless citizenship. These five ideal types show different manifestations around the way digital citizens react to an increasing postpandemic datafication processes (dataism; Lohr 2015). As such, these ideal types certainly trace emancipatory strategies previously outlined—blockchain decentralized architecture, DAOs, and data co-operatives—which serve as pillars of this resistance. They provide tangible mechanisms through which individuals and communities can reclaim ownership, reshape power dynamics, and challenge the status quo. By leveraging these strategies, utopian resistance transcends mere critique, offering tangible alternatives that realign the digital landscape with human values and aspirations.

In conclusion, as we navigate the intricate path of data (un)sustainability, these regimes provide a navigational compass, guiding us towards solutions that harness the power of data while safeguarding individual rights and societal wellbeing. By understanding and adapting to these new paradigms of citizenship, we can forge a future where data serves as a catalyst for positive change and human flourishing.

Emancipatory datafication strategies

In the labyrinth of data (un)sustainability around the geopolitical global order (Khanna 2016; Loukissas 2019), the quest for emancipation takes centre stage (Calzada and Cobo 2015). As we navigate the entangled threads of technological power and societal impact, three distinct avenues stand out as potent emancipatory strategies (Khan et al. 2022; Löhr 2023): (i) blockchain decentralized architecture, (ii) decentralized autonomous organizations (DAOs), and (iii) data co-operatives

(Calzada 2023a, 2023b). Each of these strategies offers a unique lens through which to reimagine the data landscape, empowering individuals and communities to reclaim ownership, agency, and control.

i Blockchain Decentralized Architecture

Emerging and disruptive technologies, such as blockchain, promise to disrupt conventional paradigms of data ownership and control. Matthew Zook's work on the potential of blockchain underscores its capacity to democratize data access and enhance transparency (2023). However, this potential must be harnessed responsibly, considering both its potential benefits and pitfalls. Shoshana Zuboff's groundbreaking concept of 'Surveillance Capitalism' exposes the insidious ways in which our data is commodified and exploited for profit (2019). The clash between this data-driven capitalism and the principles of privacy and autonomy raises questions about the true costs of our digital existence. Blockchain, often associated with cryptocurrencies, harbours the potential to revolutionize data management and ownership. Its decentralized architecture, secured through cryptographic techniques, ensures transparency and immutability—a stark departure from the centralization and opacity characterizing traditional data systems. By utilizing blockchain, data can be stored and accessed in a distributed manner, mitigating the risks of single points of failure and unauthorized access. Blockchain's immutability ensures that once data is entered into the system, it cannot be altered or deleted without consensus from the network. This feature holds immense promise for preserving the integrity of records, making it a powerful tool in contexts ranging from supply chain management to healthcare records. By incorporating blockchain into datafication emancipatory strategies, individuals can retain ownership of their data, enabling them to selectively share it with trusted parties while maintaining control over its use.

ii Decentralized Autonomous Organizations

The emergence of DAOs presents a radical reimagining of organizational structures and decision-making processes. DAOs operate on blockchain networks, allowing participants to collectively make decisions through a transparent and consensus-based approach. This decentralization extends to governance, resource allocation, and strategic planning. DAOs introduce the concept of 'code is law,' meaning that the pre-programmed rules embedded in the system dictate operations, bypassing intermediaries and hierarchical structures. In the context of datafication emancipatory strategies, DAOs can serve as frameworks for community-driven data governance. By enabling individuals to collectively determine data usage policies, these organizations shift power away from tech conglomerates and empower communities to define their digital destinies.

iii Data Co-operatives

Data Co-operatives embody the concept of collective ownership and control over data resources (Bühler et al. 2023b; Calzada 2021a). These cooperatives are formed by individuals or organizations who pool their data to create shared resources. This model offers an alternative to the current data economy, where vast amounts of personal information are harvested for profit without adequate compensation to the data sources. Data co-operatives aim to democratize data ownership by allowing members to collectively negotiate data transactions with external entities. This approach ensures that the benefits of data utilization are equitably distributed among contributors, fostering a sense of agency and economic empowerment. By adopting data co-operatives within datafication emancipatory strategies, individuals can regain control over their data's destiny while promoting fairness and cooperation. In a world grappling driven by artificial intelligence (AI) with the intricate dance of data's potential and perils (Helberger and Diaskopoulous 2023; Kim et al. 2018; van Noordt and Tangi 2023), these emancipatory strategies provide tangible pathways forward.

Blockchain's decentralized architecture lays the foundation for secure and transparent data management, while DAOs introduce a revolutionary approach to decision-making and governance. Data co-operatives bridge the gap between collective ownership and fair compensation, enabling individuals to harness the value of their data without falling prey to the excesses of surveillance capitalism. As we stand at the crossroads of possibility, the convergence of these strategies signals a shift towards a more equitable and sustainable data landscape. By weaving together, the threads of blockchain, DAOs, and data co-operatives, we craft a narrative of empowerment, resilience, and liberation—a narrative that not only counters the looming shadow of big tech's dominance but also paves the way for a future where data serves as a force for positive change, collective progress, and individual autonomy.

Final remarks

In the intricate tapestry of data (un)sustainability, our journey has taken us through the visionary aspirations of John Perry Barlow, the cautionary tales of Bridle's 'Dark Side,' the digital utopias painted by Cancela and Echeverría, Newport's call for digital minimalism, O'Shea's plea for future historians, Lehdonvirta's analysis of cloud empires, Zook's exploration of blockchain, Zuboff's critique of surveillance capitalism, Sadowski's unveiling of datafication, and Calzada's investigation of emerging digital citizenship regimes. We've delved into the strategies of (i) blockchain, (ii) DAOs, and (iii) data co-operatives, uncovering pathways towards emancipation from the clutches of data extractivism (Calzada 2023a, 2023b).

Our exploration of (i) blockchain's decentralized architecture, (ii) DAOs, and (iii) data co-operatives reveals a convergence of strategies that could pave the way for a more sustainable and equitable data ecosystem. By integrating these strategies, we can forge a path towards a future where individuals regain control over their data, communities shape their digital destinies, and technology is harnessed to serve humanity's collective aspirations. Yet, realizing this vision demands a grounded understanding of the complexities involved.

However, addressing the significant issue of big tech's dominance requires more than just technological solutions—it demands collaborative efforts from governments, regulatory bodies, and civil society. A potential approach could involve establishing regulatory frameworks that incentivize the adoption of decentralized technologies and data co-operatives. By encouraging tech giants to transition towards more democratic and equitable data practices, we can foster an ecosystem where competition thrives and data extractivism diminishes (Calzada 2023a).

In conclusion, the odyssey through data (un)sustainability has illuminated the potential for change. Through a synergy of blockchain, DAOs, and data co-operatives, a future beckons where data is not a tool of exploitation but a catalyst for empowerment. As we embark on new research lines, we must remain realistic about the challenges that lie ahead. Navigating the transition from data extractivism to emancipation requires a concerted effort involving multi-stakeholder policy frameworks (Calzada 2023b). By embracing these strategies and advocating for systemic change, we can reshape the digital landscape, restore agency to individuals and communities, and pave the way for a data ecosystem that reflects the ideals of true technological progress—one that uplifts humanity rather than subjugates it.

References

Arendt, H. (1966). *The Origins of Tatalitarism*. London: Penguin.

Barlow, J.P. (1996). A declaration of the independence of cyberspace. Available at: https://vimeo.com/111576518?ref=tw-v-share (Accessed 1 November 2023).

Bauwens, M., Kostakis, V., & Pazaitis, A. (2019). *Peer to Peer: The Commons Manifesto*. London: University of Westminster Press.

Bignami, F., Calzada, I., Hanakata, N., & Tomasello, F. (2022), Data-driven citizenship regimes in contemporary urban scenarios: An introduction. *Citizenship Studies*, 27(2), 145–159. doi: 10.1080/13621025.2022.2147262.

Bridle, J. (2019). *New Dark Age: Technology and the End of the Future*. London: Verso.

Bucher, T. (2012). Want to be on top? Algorithmic power and the threat of invisibility on Facebook. *New Media & Society*, 14(7), 1164–1180. doi: 10.1177/1461444812440159

Bühler, M.M., Calzada, I., Cane, I., Jelinek, T., Kapoor, A., Mannan, M., Mehta, S., Mookerje, V., Nübel, K., Pentland, A., Scholz, T., Siddarth, D., Tait, J., Vaitla, B., & Zhu, J. (2023a). Unlocking the power of digital commons: Data cooperatives as a pathway for data sovereign, innovative and equitable digital communities. *Digital*, 3(3), 146–171. doi: 10.3390/digital3030011.

Bühler, M., Calzada, I., Cane, I., Jelinek, T., Kapoor, A., Mannan, M., Mehta, S., Mookerjee, V.S., Nübel, K., Pentland, A., Scholz, T., Siddarth, D., Tait, J., Vaitla, B., & Zhu, J. (2023b). Harnessing digital federation platforms and data cooperatives to empower SMEs and local communities. TF-2: Our common digital future: Affordable, Accessible, and inclusive digital public infrastructure. *G20/T20 Policy Brief*. doi: 10.13140/RG.2.2.22347.98083/1.

Burr, C. & Floridi, L. (2020). The ethics of digital well-being: A multidisciplinary perspective. In Burr, C., Floridi, L. (Eds.), *Ethics of Digital Well-Being. Philosophical Studies Series*. Cham: Springer. doi: 10.1007/978-3-030-50585-1_1.

Burrell, J. & Fourcade, M. (2021). The society of algorithms. *Annual Review of Sociology*, 47(1), 213–237. doi: 10.1146/annurev-soc-090820-020800.

Buterin, V. (2022). *Proof of Stake: The Making of Ethereum and the Philosophy of Blockchains*. New York: Seven Stories.

Cancela, E. (2023). *Utopías Digitales: Imaginar el Fin del Capitalismo*. London: Verso.

Calzada, I. (2023a). Blockchain-driven digital nomadism in the basque e-diaspora. *Globalizations*. doi: 10.1080/14747731.2023.2271216.

Calzada, I. (2023b). Disruptive technologies for e-diasporas: Blockchain, DAOs, data cooperatives, metaverse, and ChatGPT, *Futures*, 154, 103258, doi: 10.1016/j.futures.2023.103258.

Calzada, I. (2022). *Emerging Digital Citizenship Regimes: Postpandemic Technopolitical Democracies*. Bingley: Emerald. ISBN: 9781803823324. doi: 10.1108/9781803823317.

Calzada, I. (2021b). The Right to have digital rights in smart cities. *Sustainability*, 13(20), 11438. doi: 10.3390/su132011438.

Calzada, I. (2021a). Data co-operatives through data sovereignty. *Smart Cities*, 4(3), 1158–1172. doi: 10.3390/smartcities4030062.

Calzada, I. (2018). (Smart) citizens from data providers to decision-makers? The case study of Barcelona. *Sustainability*, 10(9), 3252. doi: 10.3390/su10093252.

Calzada, I. & Cobo, C. (2015). Unplugging: Deconstructing the smart city. *Journal of Urban Technology*, 22(1), 23–43. doi: 10.1080/10630732.2014.971535.

Cheney-Lippold, J. (2016). Jus Algoritmi: How the national security agency remade citizenship. *International Journal of Communication*, 10, 1721–1742.

Couldry, N. & Mejias, U. (2019). *The Costs of Connection: How Data Is Colonizing Human Life and Appropriating It for Capitalism*. Palo Alto: Stanford University Press.

De Filippi, P. & Schingler, J.K. (2023). *Coordi-Nations: A New Institutional Structure for Global Cooperation*. Available at: https://jessykate.medium.com/coordi-nations-a-new-institutional-structure-for-global-cooperation-3ef38d6e2cfa (Accessed 1 November 2023).

Dupont, Q. (2017). Experiments in algorithmic governance: A history and ethnography of 'The DAO,' a failed decentralized autonomous organisation. In Campbell-Verduyn, M. *Bitcoin and Beyond: Cryptocurrencies, Blockchains, and Global Governance*. Milton: Routledge, pp. 157–177.

Echeverría, J. (1994). *Telépolis*. Barcelona: Destino.

European Commission. (2020). *Proposal for a regulation of the European Parliament and the Council on Contestable and Fair Markets in the Digital Sector (Digital Market Act)*. Luxembourg: European Commission.

Farrell, E., Minghini, M., Kotsev, A., Soler-Garrido, J., Tapsall, B., Micheli, M., Posada, M., Signorelli, S., Tartaro, A., Bernal, J., Vespe, M., Di Leo, M., Carballa-Smichowski, B., Smith, R., Schade, S., Pogorzelska K., Gabrielli, L., & De Marchi, D. (2023). *European Data Spaces: Scientific Insights into Data Sharing and Utilisation at Scale*. Luxembourg: Publications Office of the European Union. doi: 10.2760/400188, JRC129900.

Forestal, J. (2020). Constructing digital democracies: Facebook, arendt, and the politics of design. *Political Studies*, 69, 26–44. doi: 10.1177/0032321719890807.

Gorwa, R. (2019). What is platform governance? *Information, Communication & Society*, 22(6), 854–871. doi: 10.1080/1369118X.2019.1573914.

Hall, A. (2022). *What the History of Democracy Can Teach Us About Blockchain Governance*. Available at: https://thedefiant.io/what-the-history-of-democracy-can-teach-us-about-blockchain-governance (Accessed 1 November 2023).

Helberger, N. & Diakopoulos, N. (2023). ChatGPT and the AI act. *Internet Policy Review*, 12(1). doi: 10.14763/2023.1.1682
Hubbard, S. (2023). *Decentralized Autonomous Organizations and Policy Considerations in the United States*. Harvard: Belfer Center.
Hughes, L., Dwivedi, Y.K., & Misra, S.K. (2019). Blockchain research, practice, and policy: Applications, benefits, limitations, emerging research themes and research agenda. *International Journal of International Management*, 49, 114–129. doi: 10.1016/j.ijinfomgt.2019.02.005.
Isin, E. & Ruppert, E. (2015). *Being Digital Citizens*. NYC: Rowman & Littlefield.
Inwood, O. & Zappavigna, M. (2021). Ideology, attitudinal positioning, and the blockchain: A social semiotic approach to understanding the values construed in the whitepapers of blockchain start-ups. *Social Semiotics*, 0(0), 1–19. doi: 10.1080/10350330.2021.1877995.
Khan, K., Su, C.-W., Umar, M., & Zhang, W. (2022). Geopolitics of technology: A new battleground? *Technological and Economic Development of Economy*, 28(2), 442–462. doi: 10.3846/tede.2022.16028.
Khanna, P. (2016). *Connectography: Mapping the Global Network Revolution*. NYC: Weidenfeld & Nicholson.
Kim, Y.M., Hsu, J., Neiman, D., Kou, C., Bankston, L., Kim, S.Y., Heinrich, R., Baragwanath, R., & Raskutti, G. (2018). The stealth media? Groups and targets behind divisive issue campaigns on Facebook. *Political Communication*, 35, 515–541. doi: 10.1080/10584609.2018.1476425.
Lehdonvirta, V. (2022). *Cloud Empires: How Digital Platforms Are Overtaking the State and How We Can Regain Control*. Boston: MIT Press.
Lohr, S. (2015). *Data-ism: The Revolution Transforming Decision Making, Consumer Behavior, and Almost Everything Else*. NYC: OneWorld.
Löhr, G. (2023). Conceptual disruption and 21st century technologies: A framework. *Technology in Society*, 74, 102327. doi: 10.1016/j.techsoc.2023.102327.
Loukissas, Y. A. (2019). *All Data Are Local: Thinking Critically in a Data-Driven Society*. Boston: MIT Press.
Marquardt, F. (2021). *The New Nomads: How the Migration Revolution Is Making the World a Better Place*. NYC: Simon & Schuster.
Mathew, A.J. (2016). The myth of the decentralised internet. *Internet Policy Review*, 9(3). Available at: https://policyreview.info/articles/analysis/myth-decentralised-internet.
Monsees, L. (2019). *Crypto-Politics: Encryption and Democratic Practices in the Digital Era*. Oxon: Routledge.
Morozov, E. (2022). Critique of techno-feudal reason. *New Left Review*, 133–134, 89–126.
Morozov, E. & Cancela, E. (2023). *Benedetta Brevini Interview*. Available at: https://www.eldiario.es/tecnologia/benedetta-brevini-grandes-tecnologicas-dicen-no-emitir-carbono-inteligencia-artificial-ayudar-petroleras_128_10240272.html (Accessed 1 November 2023).
Nabben, K. (2022). A political history of DAOs. Available at: https://www.fwb.help/wip/cypherpunks-to-social-daos (Accessed 1 November 2023).
Newport, C. (2019). *Digital Minimalism: Choosing a Focused Life in a Noisy World*. London: Penguin.
O'Shea, L. (2021). *Future Histories: What Ada Lovelace, Tom Paine, and the Paris Commune Can Teach Us About Digital Technology*. London: Verso.
Polanyi, K. (1944). *The Great Transformation: The Political and Economic Origins of Our Time*. London: Beacon Press.

Rennie, E., Zargham, M., Tan, J., Miller, L., Abbott, J., Nabben, K., & De Filippi, P. (2022). Towards a participatory digital ethnography of blockchain governance. *Qualitative Inquiry*, 28(7), 837–847. doi: 10.1177/10778004221097056.

Sadowski, J. (2019). When data is capital: Datafication, accumulation, and extraction. *Big Data & Society*, 6(1), 1–12. doi: 10.1177/2053951718820549.

Schumpeter, J.A. (1942). *Capitalism, Socialism, and Democracy*. NYC: Harper & Brothers.

Spelliscy, C., Hubbard, S., Schneider, N., & Vance-Law, S. (2023). *Toward Equitable Ownership and Governance in the Digital Public Sphere*. Harvard: Belfer Center.

Srivastava, S. (2021). *Algorithmic Governance and the International Politics of Big Tech*. Cambridge: Cambridge University Press.

Stanford DAO Workshops 2022 and 2023 (2023). *Frances C. Arrillaga Alumni Center at Stanford University. DAO Research Collective, Megagov, Smart Contract Research Forum, and Stanford Center for Blockchain Research*. Available at: https://m.youtube.com/watch?v=mTT0tCix43E.

Stucke, M. E. (2022). *Breaking Away: How to Regain Control Over Our Data, Privacy, and Autonomy*. New York: Oxford University Press. doi: 10.1093/oso/9780197617601.003.0001.

Taplin, J. (2017). *Move Fast and Break Things: How Facebook, Google, and Amazon Have Cornered Culture and What It Means for All Of Us*. NYC: Little Brown.

Toscano, J. (2021). *Data Privacy Issues Are the Root of Our Big Tech monopoly Dilemma. Forbes*. Available at: https://www.forbes.com/sites/joetoscano1/2021/12/01/data-privacy-issues-are-the-root-of-our-big-tech-monopoly-dilemma/?sh=4be10acc3cfd (Accessed 1 November 2023).

Van Dijck, J. (2018). *The Platform Society: Public Values in a Connective World*. Oxford: Oxford University Press.

Van Noordt, C. & Tangi, L. (2023). The dynamics of AI capability and its influence on public value creation of AI within public administration. *Government Information Quarterly*, 101860. doi: 10.1016/j.giq.2023.101860.

Veliz, C. (2020). *Privacy is Power: Why and How You Should Take Back Control of Your Data*. London: Penguin.

Viano, C., Avanzo, S., Boella, G., Schifanella, C., & Giorgino, V. (2023). Civic blockchain: Making blockchains accessible for social collaborative economies. *Journal of Responsible Technology*. doi: 10.1016/j.jrt.2023.100066.

WEF. (2023). *Decentralized Autonomous Organization (DAO) Toolkit*. Davos: WEF.

Zook, M. (2023). Platforms, blockchains and the challenges of decentralization. *Cambridge Journal of Regions, Economy and Society*, XX, 1–6. doi: 101093/cjres/rsad008.

Zuboff, S. (2019). *The Age of Surveillance Capitalism: The Fight for a Human Future at the New Frontier of Power*. NYC: Profile.

10

EMBEDDING SUSTAINABILITY IN SOFTWARE DESIGN AND DEVELOPMENT

Accessible digital tools for local communities

Cristina Viano, Guido Boella, and Claudio Schifanella

The challenge

According to some news commentators, social media is becoming a lot less social (Chen 2023). We have moved from communicating with connected persons in earlier social networks to broadcasting to a wider audience in modern social media (Bogost 2022). Some users have reacted to this development by looking for smaller, themed social networks to restore a sense of community. Along the same lines, there is a growing demand for more education on responsible use of social media instead of imposing oversimplified restrictions (Nesi 2023; OECD 2023). While something is changing in the way established digital platforms are perceived, newer technologies (like blockchain) are under scrutiny for their social and environmental impact. Several news articles in 2022 reported on speculative bubbles and hacks in cryptofinance (see for instance Korn (2022) and Sherman (2022)). Web-based think-tanks such as The Crypto Syllabus have started a critical conversation about crypto worlds and their (geo)political and economic effects (Morozov 2023). It is more difficult to find information about alternative blockchain experimentations, such as the ongoing experiments of the Kolectivo impact network in Curaçao and Circles basic income in Berlin.

These news are illustrative of the promises and pitfalls of digital technologies as apparent solutions for contemporary socio-economic and environmental problems. There is an increasing demand for alternative digital tools that address the inequalities perpetuated by digital capitalism models.

Alternative models and concrete experimentations are advanced by scholars, activists, and digital social innovators. These are based on different discourses about the relationship between digitalisation and sustainability. Many of these actors see

DOI: 10.4324/9781003441311-12

digitalisation as an opportunity for transformation towards non-growth-based, post-capitalist socio-economic orders; but they are also concerned about their ecological risks. Others aim to reform and regulate market economies, and see digitalisation as balancing growth and sustainable lifestyles (Adloff and Neckel 2019; Lenz 2021).

The first category includes approaches based on concepts such as *convivial technologies* (Vetter 2018) and *low technologies* (Bihouix 2014). They explore the possibilities of hardware and software based on short supply chains, low environmental impact, collaborative development, and open-source access. Examples include the Free/Libre Open Source Software and Platform Cooperativism movements, the Small Tech initiative (Balkan 2019), the Soft Digitalization framework (Lange and Santarius 2020), and the Mid Tech proposal (Kostakis et al. 2023). Other approaches, such as Slow Technology from the field of design (Hallnas 2015), are focused on the user experience, having in mind wellbeing rather than just efficiency. Slow Technology is also recognized in the field of computer ethics as a broaderdesign approach forInformation and Communication Technologies (ICTs), prioritizing human-centred, environmentally sustainable, and socially just practices (Patrignani and Whitehouse 2014). Many civil society initiatives are inspired by the idea of "slow" digital solutions in different socio-economic sectors including democratic participation, civic activism, and social and solidarity economies – see for instance Olivier and Wright's (2015) work on *digital civics*.

The first question is then: *how can the general principles advanced by the "slow" and "soft" approaches to digital processes be put in practice to develop sustainable technologies for civic participation and social economies?*

Indeed, the actual opportunities to develop and choose alternative technologies are constrained by the pervasiveness of big tech monopolies. Moreover, many initiatives still require that to use open-source tools, participants must have medium-to-high-level technical skills and a strong awareness of their potentialities and critical points.

What, then, are the options for researchers in computer science and digital studies developing technologies that seek to overcome these barriers?

This chapter presents the approach of the Digital Territories and Communities research group at the Computer Science Department of the University of Turin. This contribution reflects upon the strategy of a computer science-based but interdisciplinary research group, aiming to situating that strategy within the wider context of current debates in the digital geography community and encouraging dialogue on these topics. These debates are about: how to overcome binary representations of digital and non-digital domains (McLean 2020), the embeddedness of local participatory platforms (Mello Rose 2021), and how digital social innovation can support emancipatory practices in urban communities (Certomà 2021). We also refer to recent works in geography (McLean 2020), and environmental and media studies (Kuntsman & Rattle 2019; Lenz 2021) on the *environmental* aspects of digital sustainability, adapting their analytical frameworks to shed light on the *social* sustainability of digitalisation. This latter aspect has been the core focus of the research group so far.

The digital territories and communities research group

Aims and digital tools

The Digital Territories and Communities research group was founded by computer scientists in 2017 to develop digital tools that enable civic and participatory practices in urban communities. The aim is to bring the advantages of global mainstream technologies to the local communities and direct them towards civic and public goals.

Our experimental research started with FirstLife, a geolocated civic social network based on an interactive map (Boella et al. 2019). FirstLife is a tool for community members to support coordination and cooperation around initiatives of public interest in local areas (e.g. a neighbourhood). On the map and in the newsfeed, users can create their own content (events, news, proposals, stories, groups) relating to their experience of the local area, and openly interact through chats, polls, and file uploads (Figure 10.1).[1]

The second tool, CommonsHood, is a blockchain-based wallet app aimed at strengthening social collaborative economies (Balbo et al. 2020). It supports exchange models such as community currencies, libraries of things, and rewards for contributing to urban commoning. Any user, without any knowledge of blockchain or coding, can create and exchange cryptographic tokens[2] representing relevant values within local communities: coins, coupons, digital certificates, access to shared resources and spaces, etc. (Figure 10.2).[3]

The two platforms have been co-designed and are tested and iteratively developed in different local contexts via participatory research and living labs methodologies (Ballon and Shurman, 2015), working with local authorities, schools, civil society organisations, and local economic actors.

Interdisciplinarity

To deal with the complexity of socio-technical processes, the research group has opened up to other disciplines: computer ethics, economics, economic sociology, urban studies, geography, and pedagogy. Moreover, professional expertise is combined to help ground the experimentations in their local social contexts via participatory design, community development, social policy, and diversity management.

It has recently been noted that future research on digital sustainability should look at "how digitalization and sustainability are not only discursively mediated or criticized, but also how they are implemented at the level of practical and political actions" (Lenz 2021: 202). Focus should also be directed towards "the everyday

FIGURE 10.1 The FirsLife interface: newsfeed and map.

Source: Created by authors.

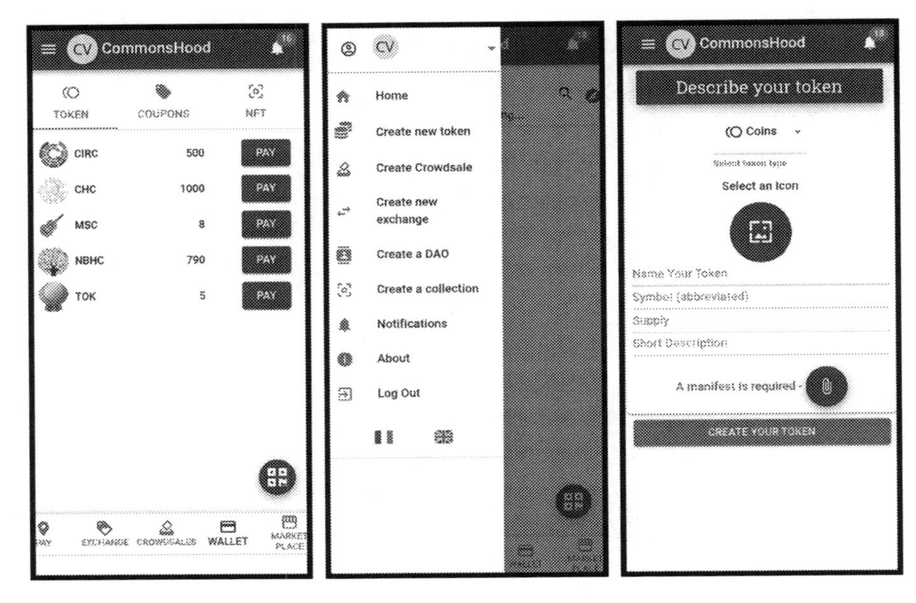

FIGURE 10.2 The CommonsHood interface: wallet, main menu, token creation functionality.

Source: Created by authors.

practice of people" (ibid) and their motivation for engaging with digital activities (McLean 2020). Our research group aims to fill this gap by stimulating concrete actions of digital social innovation.

At a more general level, our interdisciplinary work is based on two assumptions. The first assumption is that social and technical sciences enhance one another. Attention is paid not only to complementing technical developments with social research, thereby tailoring technologies to better meet social needs, but also fostering technical awareness within the social sciences studying digital transformations (Ash et al. 2016; Iapaolo et al. 2023). The second assumption is that technologies are not neutral: their value resides not only in their purpose but also in all their design, development and implementation processes (Certomà 2021), and the design phase is critical because socio-political visions are encoded in their technical features (Iapaolo et al. 2023).

The social sustainability of the digital

Sustainability is another ubiquitous but contested concept (Adloff and Neckel 2019). Referring to the sustainability of local and urban communities, Agyeman presented two challenges (Agyeman 2012). The first is the *scientisation of sustainability*: scientifically, we know what to do, and how, to achieve sustainability goals, but the challenge concerns the social sciences and a shift in the political and civic cultures. The second challenge is *the need to foreground issues of equity and social justice* in most "green" approaches, for which he advances the concept of just sustainabilities (Agyeman et al. 2003).

Such challenges are all relevant to more specific issues related to digital sustainability. We could, to some extent, temper Agyeman's claim that scientifically we already know what to do and how, since developments in digital technologies happen at an extremely rapid pace, which constantly poses new challenges. However, it is still true that the major challenge is cultural and political. On the other hand, the pervasiveness of digitalisation in our lives brings both opportunities and new risks for equity and justice. Hence, the motivation to undertake research in software development that activates social innovation, guided by social aims.

As regards the *social* side of digital sustainability, there are tensions in five interrelated areas: (i) the business models of *digital monopolies* and their implications for *digital data ownership*; (ii) *digital relationships,* that expand opportunities but also risk commodifying and automating social and personal interactions; (iii) the *digital spaces* generated by the platforms that are simultaneously embedded and disembedded from local contexts; (iv) new socio-spatial inequalities and digital divides that threaten *digital justice*; and (v) *digital transformations in local communities.*

Van Dijk, Poell, and De Waal (2018) have observed that actors involved with the *platform society* are increasingly concerned with safety, transparency, privacy, and willingness to protect individual interests. On the other hand, the public and collective values of fairness and social responsibility are less regarded. We describe how our approach to digitalisation can contribute to steering digital platforms from individual interests towards public goods.

The digital territories and communities approach

Our understanding of social digital sustainability takes the form of five principles that guide our digital design and development. These principles address the five areas of tensions mentioned above.

Re-deploying mainstream technologies

Big tech platforms provide low-cost and efficient services, making it difficult to counter their (quasi-)monopolistic positions. Radically new architectural solutions often take a long time to establish themselves due to the limited network effect (Adloff and Neckel 2019). To overcome this, mainstream solutions can be re-deployed, re-designed from the outset as open access and open source, and oriented towards local implementation. However, taking inspiration from technologies that are well-established (e.g. social networks) or hyped (e.g. blockchain) does not mean uncritically accepting their narratives. Rather, we should be asking which social goals and public values their affordances can serve and directing their re-design accordingly.

In the project *Riscopri Risorse*, (meaning Rediscovering Resources), the First-Life platform was used by middle-school students to map run-down urban spaces and discuss proposals for their restoration. Micro-interventions, such as urban graffiti, were then implemented. The research team encouraged the students to reflect on the particularities of the social media tool, which integrates GIS functionalities with social network functionalities (chat, sharing images, etc.). Contents on the map can be posted and interactively updated by everyone, which makes it quite different to Google Maps (where users can only leave reviews) and more static crowd mapping tools. Connections are nurtured among people who do not necessarily know each other, offering a more social interaction than being restricted to close bubbles. The content must be of public interest, and students were guided to write proposals and opinions in a constructive and thoughtful way, as opposed to the immediate outpouring of feelings encouraged by mainstream social networks.

The core innovation brought by the blockchain is the possibility to digitally represent assets (e.g. currency, data, assets, certificates etc.) in the form of tokens, and to implement secure and transparent transactions of such tokens, without relying on intermediaries for guaranteeing the authenticity. Moreover, smart contracts allow to programme the transactions. As for the CommonsHood wallet app, the properties mentioned above are not leveraged for speculative purposes like many cryptocurrencies. On the contrary, they are re-interpreted to provide local social economies with the necessary tools to make (im)material assets more liquid and to implement schemes such as crowdfunding, group buying, barter circuits, proximity marketing and new forms of the sharing economy. As opposed to the trustless trust (Werbach 2018) approach of cryptocurrencies, which contrasts with solidarity and community principles, the possibility to avoid intermediaries that the blockchain allows is leveraged to encourage socio-economic exchange among different stakeholders that would not otherwise cooperate (e.g., by making non-monetary values liquid through tokens), to lower the cost of commercial services (enabling the autonomous creation of discount coupons and crowdfunding mechanisms), or to create purpose-driven tokens to reward community-oriented behaviours. The strategy differs from that of projects that develop new blockchains based on social economy principles at the architecture level (e.g. HoloChain), although it does bring the risk of limited popularity and maintenance. In fact, the tokenisation tools offered by CommonsHood are based on standard smart contracts like ERC20 and ERC721. Specific social economy features are built on top of such contracts.

The business models of social networks and most other web applications and apps are based on exploiting data extracted from users, as extensively discussed by Zuboff (2019). This has ethical and legal repercussions for data ownership, privacy, surveillance and user remuneration. It also leads to the trend of keeping users on the apps at all costs, exposing them to personalised advertisements, possibly tainted by biases. "Addictive" emotive content is spread (e.g. hate speech, misinformation), which impacts on public discourse. In order to foster users' ownership of data, we have chosen to avoid profiling users based on the extraction of personal data and biased argumentation algorithms. In FirstLife, the information is filtered only by the locality dimension.

Making digital technologies civic

Digital relationships mediated by platforms are often based on individual interests and incentives. The ownership, business and communication models of mainstream digital platforms can clash with the emancipatory principles of social economies (Krlev et al. 2021) and civic initiatives.

On the other hand, the digital technologies we develop are aimed at supporting participatory practices and collective goals, which we define as *civic*. From a governance perspective, this means that civil society, small-scale economic actors, and local institutions interact in various configurations, in the co-production of services or self-organisation of community members. From an economic perspective, we refer to socio-economic microtransactions marked by both collaborative/solidarity logic and market exchanges. Hence, the proposal of *civic social networks* and *civic blockchains*. Even though the term *civic technologies* remains subject to debate (Certomà and Corsini 2021), we use it to stress that these technologies are designed to reshape systems of governance and power (Shrock 2019), rather than simply being used for collective and public ends.

New solidarities and a mix of social and economic incentives are exemplified by the Collegno Local Lab, activated within the EC-funded project NLAB-4CIT.[4] Collegno, a municipality in northern Italy with 30,000 inhabitants, serves as a testing ground for CommonsHood. Here, young volunteers receive "reward tokens" in their wallet, issued by the municipality. Tokens can be exchanged with coupons that provide benefits in local commercial and cultural services. The purpose is to stimulate new forms of interactions among actors that would not cooperate otherwise (e.g., giving a teenager access to cultural services and local shops currently threatened by mass distribution and e-commerce), rather than monetise volunteering activities. Of course, in every tokenisation mechanism, there is the risk of monetising social relationships. Therefore, careful social design is needed. In Collegno, local actors decided to distribute tokens to volunteers in proportion to the time they dedicated, but not based on strict accounting of the single hours. This way, "small" commitments are given value as well, and accumulation mechanisms are not encouraged. Participants can also donate their tokens to friends.

Making digital technologies local

Mainstream urban platforms are simultaneously embedded and disembedded from the local geographies they mediate. They depend on local resources and reshape local socio-economic interactions, but they can also extract value (Graham 2020) without necessarily compensating those who generate it.

An essential component of our approach is the relationship between the digital platforms and their geographical contexts of usage: small cities, neighbourhoods, or urban commons.

Firstly, digital interactions are meant to not only complement but also encourage, rather than discourage, physical encounters in urban spaces, and to stimulate collective initiatives around material commons or public spaces.

In the EC-funded research project CO3,[5] the concept of *augmented commoning areas (ACA)* was developed to model urban commons (here broadly meant as both public and community spaces such as gardens, social hubs, community centres, etc.) where community members can cooperate through specific digitally enabled interactions. A mobile phone application enabled such interactions by integrating various technologies: FirstLife's *geolocated map* and *augmented reality* visualisation were used to find tasks and opportunities in the ACA provided by local associations; a *blockchain wallet* was used for exchanging cryptographic tokens as a community currency; and an *interactive democracy* tool was employed for collective decision-making. An example ACA is the following. A neighbourhood community centre issues its own local coins in the blockchain wallet. Coins are sold to citizens as prepaid cards or distributed as rewards. A theatre company, active in the centre, prepares a coupon to invite people to subscribe to its acting course with a discount. Sarah, a visitor to the ACA, thanks to AR functionalities, visualises the virtual notice board and discovers the coupon. She books the course using the house coins. She receives a reward badge, which motivates here to buy and drop a "service on-hold" (a coffee or a meal at the centre's restaurant) to be offered for free to someone in need. She also participates to an online poll, advertised on the geolocated map, for suggestions on the next events to be organised by the company theatre. Note that in the initial design, to encourage in-person interactions, the activation of AR, blockchain, and e-voting functionalities was contingent upon geolocated proof of presence in the ACA. The COVID-19 pandemic forced developers to also allow remote use. Interestingly, participants did not wish to return to remote interactions as the only or main mode of interaction afterwards.

The second aspect of the relationship between the digital platforms and their geographical contexts is that FirstLife and CommonsHood are expected to be adaptable to different local contexts, and shaped by the local community's needs in terms of functionalities and contents.

With both platforms, a core set of *functionalities* is provided in a modular structure so that each community can choose what to activate. In most cases, the functionalities are customised or new functionalities are added. Indeed, on one side, this approach does not mean developing an application from scratch, based on the community requests. However, the two tools offered are not limited to a set of predefined functionalities. Finding a balance between context-specific requirements (that also evolve over time, as do social needs) and a stable core structure is a major challenge. Users' high expectations of flexibility can conflict with technical constraints.

As regards the *contents*, the decisions of local actors about what to share on the two apps are crucial for their relevance to local needs. This can seem self-evident for FirstLife, being a map-based social network in which users necessarily relate content to specific places. Conversely, in CommonsHood, the fact that the choice of values to tokenise is up to the local community makes the app different to all other blockchain wallet, including those designed for social economies. Such flexibility requires significant effort in the preliminary co-design of the socio-economic models to be implemented, which also raises the question of which relevant actors should have a voice in the co-design, and how this affects future inclusion and exclusion processes. Example of these negotiations are provided by the discussion that took place in the Collegno Local Lab (see above Section 4.2) on the value and usage rules of the reward tokens. Educators and youth representatives were incisive in prioritising the social value of the tokens over the economic one, avoiding commodification mechanisms and encouraging the technical possibility that tokens are donated to others' wallet. Conversely, local retailers were less interested in co-designing the process.

Making digital technologies accessible

Among the many things expected of digitalisation is democratising access to information, services, and decision-making processes, which is meant to have positive effects on social justice and equity. However, economic, cultural, and/or technical digital divides can fundamentally undermine accessibility to hardware and software. Meanwhile, digital processes can disempower or reproduce inequalities (Certomà 2021; Robinson 2015). We address these from two perspectives, conscious of the fact that they are entry points to complex problems.

The first perspective concerns the *accessibility of the digital tools and processes themselves*. Here, we refer to accessibility in the general sense: making the digital tools available to as many people as possible.[6]

First, we aim to achieve technical accessibility by making the particular affordances of the new technologies (i) available to users without specific technical skills thanks to the no-code approach (Caballar 2020), and (ii) understandable thanks to the adaptation of metaphors drawn from mainstream technologies to local contexts and civic uses. In CommonsHood, the wallet's simple interface makes the complex affordances of the blockchain (digital representations and transfer of values) accessible to users with basic digital skills, and far simpler to manage than in most other crypto wallets. The metaphor of the newsfeed in FirstLife, inspired by social networks, and the wallet in CommonsHood, inspired by crypto-wallets, are integrated with a geolocated map, so that the cartographic metaphor strengthens the local dimension.

The second solution is to work on the economic accessibility and affordability of the tools. Both platforms are web apps, accessible on any browser in a mobile phone or computer, so they do not require top-of-the-range devices. As the products of public research, they are provided open source and free of charge to local administrations, civil society organisations, and developers. However, as with many digital social innovation initiatives, the risk of exclusion remains for people in the community who do not have (or are not interested in having) access to digital devices. Whenever possible, alternative solutions are provided, such as the collective use of digital devices (e.g. when FirstLife is used in schools), the lending of devices, QR codes associated with an individual's CommonsHood wallet which allow them to receive and use tokens with the support of other wallet holders.

The second perspective on the relation between digital solutions and social justice concerns *the support that the digital tools themselves can provide for inclusive and emancipatory processes*. One example is our collaboration with the Female Toponymy association,[7] where FirstLife is used to provide educational activities on gender equality and gender urban planning. Through the map, users become aware of the small percentage of streets and urban places named after women in their city and are invited to share their proposals for naming unnamed places after women of note.

Facilitating digital transformations

The concepts of digital transition, digital transformation, and digital social innovation all refer to processes of socio-technical changes. However, social changes take place gradually and require longer time than digital innovation. Moreover, technologies alone do not guarantee transformative changes if they are not part of a broader cultural and political process. Therefore, digital facilitation is a core pillar of our approach. By this, we mean coupling technical developments with educational and community activities that activate such processes and outline common goals for the different actors. This kind of work is necessary for three main reasons.

First, there is a need to develop awareness and skills among community actors about *how* the civic digital platforms can be used and *for what purposes*. The "how" does not relate to basic digital literacy but rather to understanding the rationales of new and complex technologies. The "purposes" relate to their potential for developing social economies and civic participation, something that is not widely known while mainstream imaginaries prevail (e.g. about social networks). There needs to be a significant level of civic intentionality regarding use of the digital tools, which does not mean delegating responsibility for the sustainability of technologies to individual users (McLean 2022).

Secondly, there is a need to implement co-design processes that make it possible to adapt the platforms to local needs. The more challenging part of the work is co-designing not only the digital tools themselves but also the even earlier work on the socio-economic models they enable. As for the technical tools, digital facilitators are also interpreters and mediators between social requirements and technical developments and assistance.

The third reason for digital facilitation is the need for keeping collaborative use of the technologies alive. This is due to the nature of participatory processes, and the fact that engagement mechanisms based on algorithms and individual incentives – typical of mainstream ICTs – are intentionally avoided in the two digital platforms when they conflict with the collective and public approach.

A meaningful example of digital facilitation is CommonsHood's library of things functionality. A library of things is a micro-circular economy in which participants lend and borrow objects. These objects are represented in the app with non-fungible tokens[8] created by each lender which can be reserved via a booking system. For every action contributing to the circulation of objects, users receive other tokens representing community coins. Within the *Comunita' Organizzate per Scambio Oggetti* (COSO – Organized Communities for the Exchange of Objects) project, a small group of residents were first introduced to the (civic) blockchain and the library of things. Plenty of proposals and objections have arisen about: the community-building aims of the project, how to adapt tokenisation mechanisms accordingly, and how to change the economic lexicon of the blockchain (coins, rewards, deposits) for a lexicon more evocative of reciprocity in a community. A six-month programme of local cultural events and "offline" testing of the system allows careful design of the social mechanisms to be represented in the app's functionalities.

In the same vein, digital facilitation is the core role of our research group in the European Digital Innovation Hub on Public Administration Intelligence,[9] where social economy organisations and local authorities undertake digital social innovation processes.

The environmental sustainability of digital participatory tools

Kuntsman and Rattle (2019) identify three categories of conceptualisations about the relationship between digitalisation and environmental sustainability. Inspired by their analysis of digital communications, we reflect on digitals tools for civic participation.

These kinds of tools rarely fall under the first category of *digital as a tool* (ibid.) which is expected to enable sustainable processes directly. FirstLife and CommonsHood come under the second category of *the digital as facilitator of behavioural changes* in *education* or *sustainable consumption* processes (ibid.). FirstLife is used in schools, or in other awareness-raising projects involving young people, and by NGOs working on environmental protection.[10] CommonsHood is expected to foster sustainable consumption, but not as an e-commerce platform like many of the cases mentioned by Kuntsman and Rattle (2019). Rather, the blockchain wallet leverages tokenised incentives to encourage environmentally sustainable behaviours, as in the above-mentioned COSO project.

The third category is *the digital as a material object* (ibid.), with its impact on natural resources, energy, and e-waste. Both FirstLife and CommonsHood are web apps that do not require top-end devices. This is a first step to avoiding contributing to the continuous disposal of devices. As for energy consumption, blockchain technologies based on the proof-of-work (PoW) consensus algorithm do require considerable computational power. Fortunately, PoW is now used by very few platforms e.g. bitcoin. All popular smart-contract-based platforms, including Ethereum from September 2022, now rely on proof-of-stake (PoS) consensus algorithms, which has drastically reduced power consumption to less than 5%. CommonsHood uses a consortium blockchain, which has fewer nodes and does not require energy-consuming mining. It is based on the popular Ethereum Virtual Machine technology, and it is compatible with the European Blockchain Services Infrastructure (EBSI).

Kuntsman and Rattle adopt a critical perspective to find blind spots that foster digital solutionism. Interpreting our platforms through this lens aims to reflect on their limitations. The fostering of sustainable behaviours must not underestimate the environmental impact of the platforms themselves. We must also be willing to acknowledge when digital solutions should be diminished or avoided.

Conclusion

In this chapter, we presented our interdisciplinary approach to the development of technologies for civic purposes and its core pillars: the redeployment of mainstream technologies to make them civic and local; the focus on their technical and economic accessibility; and the leveraging of cultural processes of digital facilitation.

Examples of experimentations with a civic social network and a civic blockchain wallet show how this approach strives to find technical and cultural access points to the complexity of digital sustainability, while assuming some tensions and paradoxes.

One of these tensions is the economic sustainability of socially sustainable digital tools. Some authors advocate for institutional investments in small and independent software development projects (Balkan 2019), which often struggle to emerge. McLean affirms that governments and corporations should correct code inequities and unaccessibilities (McLean 2020). As for public institutions, we see here a significant role and responsibility for universities as well, especially computer science research groups. As such, our research group invests in applied research and makes it results available to local communities as part of its civic role.

Another tension concerns the choice to start working on the social aspects of digital sustainability, while recognising that it is no longer possible to overlook the environmental sustainability of digital infrastructure in institutional "green" strategies, including those of universities (McLean 2022).

The challenge is how to develop real technologies that are sustainable while maintaining a constructive stance that is open to the potential of digital relationships. We conclude by referring once more to the reflections of Jessica McLean (2020) and other well-known geographers she mentioned. We are inspired by Lesley Head's (2016) attitude of *hope* in the face of the Anthropocene challenges, as opposed to uncritical (technological) optimism. Moreover, Harriet Bulkeley et al. (2018) see a connection between this generative hope and urban experimentations, which is where our research fits in. Secondly, we are responding to Gibson-Graham and Roelvink's (2010) call for academic research as performative practice that activates, amplifies, connects and disseminates existing ethical community economy projects, which in our case are digitally supported or digitally enabled. All these conceptualisations are inspirational for interdisciplinary, creative and exploratory research work on digital sustainabilities.

Notes

1. https://www.firstlife.org/en/.
2. Tokens are digital representations of values and assets on a blockchain.
3. https://www.commonshood.eu/.
4. Network of Local Laboratories for Civic Technologies Coproduction.
5. CO3 – Digital Disruptive Technologies to Co-create, Co-produce and Co-manage Open Public Services along with Citizens.
6. We do not address here the topic of the accessibility of digital interfaces for specific groups of users and people with impairments, for whom design standards and good practices already exist.
7. The Italian association in Italy advocates for naming streets and urban spaces after women who have contributed to improving society.
8. Non-fungible tokens represent unique digital or material objects with a unique digital identifier.
9. https://european-digital-innovation-hubs.ec.europa.eu/edih-catalogue/pai.
10. For instance, the *Pedalè* platform for sharing proposals about bike lanes in the city of Turin (IT), the *Quartier Circolare* project for exchanges between Italian and Senegalese young environmental activists, and the *Sustainable Piedmont* platform for projects on SDGs in the Piedmont region.

References

Agyeman, J. (2012). Just Sustainabilities, September 21, 2012, available at https://julianagyeman.com/2012/09/21/just-sustainabilities/

Agyeman, J. (2013). *Introducing Just Sustainabilities. Policy Planning and Practice*. London: Zed Books.

Adloff, F. & Neckel, S. (2019). Futures of sustainability as modernization, transformation, and control: A conceptual framework. *Sustain Science*, 14, 1015–1025.

Balbo, S., Boella, G., Busacchi, P., Cordero, A., De Carne, L., Di Caro, D., Guffanti, A., Mioli, M., Sanino, A. & Schifanella, C. (2020). CommonsHood: A blockchain-based wallet app for local communities. *Presented at the Proceedings - 2020 IEEE International Conference on Decentralized Applications and Infrastructures*, DAPPS 2020, art. no. 9126008, pp. 139–144.

Balkan, A. (2019). Small technology, available at https://ar.al/2019/03/04/small-technology/

Ballon, P. & Schuurman, D. (2015). *Living Labs: Concepts, Tools and Cases*, vol. 17. London: Emerald Group Publishing Limited.

Bihouix, P. & McMahon, C. (2014). *The Age of Low Tech: Towards a Technologically Sustainable Civilization* (1st ed.). Bristol: Bristol University Press.

Boella, G., Calafiore, A., Grassi, E., Rapp, A., Sanasi, L. & Schifanella, C. (2019). Firstlife: Combining social networking and VGI to create an urban coordination and collaboration platform. *IEEE Access*, 7, 63230–63246.

Bogost, I. (2022). *The Age of Social Media Is Ending*, The Atlantic, (November 10, 2022), available at https://www.theatlantic.com/technology/archive/2022/11/twitter-facebook-social-media-decline/672074/

Bulkeley, H., Drew, G., Hobbs, R. & Head, L. (2018). Conversations with Lesley head about hope and grief in the anthropocene: Reconceptualising human-nature relations. *Geographical Research*, 56, 325–335.

Caballar, D. (2020). Programming without code: The rise of no-code software development. 11 March 2020, available at https://spectrum.ieee.org/programming-without-code-no-code-software-development

Certomà, C. (2021). *Digital Social Innovation: Spatial Imaginaries and Technological Resistances in Urban Governance*. Cham: Springer International Publishing.

Certomà, C. & Corsini, F. (2021). Digitally-enabled social innovation. Mapping discourses on an emergent social technology. *Innovation: The European Journal of Social Science Research*, 34(4), 560–584.

Chen, B.X. (2023). The future of social media is a lot less social. *The New York Times*, April 19th 2023.

Gibson-Graham, J.K. & Roelvink, G. (2010). An economic ethics for the anthropocene. *Antipode*, 41, 320–346.

Graham, M. (2020). Regulate, replicate, and resist – the conjunctural geographies of platform urbanism. *Urban Geography*, 41, 453–457.

Hallnäs, L. (2015). On the philosophy of slow technology. *Acta Universitatis Sapientiae-Social Analysis*, 5(1): 29–39.

Head, L. (2016). *Hope and Grief in the Anthropocene: Re-conceptualising Human–Nature Relations*. Abingdon and New York: Routledge.

Korn, J. (2023). Binance-linked blockchain hit by $570 million crypto theft. *CNN News*, October 7, 2022, available at https://edition.cnn.com/2022/10/07/tech/binance-bridge-hack/index.html

Kostakis, V., Niaros, V. & Giotitsas, C. (2023). Beyond global versus local: Illuminating a cosmolocal framework for convivial technology development. *Sustain Science* 18: 2309–2322.

Krlev, G., Pasi, G., Wruk, D. & Bernhard, M. (2021). Reconceptualizing the social economy. *Stanford Social Innovation Review*. https://doi.org/10.48558/98VT-G859.

Kuntsman, A. & Rattle, I. (2019). Towards a paradigmatic shift in sustainability studies. *Environmental Communication*, 13, 567–581.

Iapaolo, F., Certomà, C. & Giaccaria, P. (2023). Do digital technologies have politics?, in Osborne, T. and Jones, P. (Eds.) *A Research Agenda for Digital Geographies* (pp. 27–40), Cheltenham: Edward Elgar.

Lenz, S. (2021). Is digitalization a problem solver or a fire accelerator? Situating digital technologies in sustainability discourses. *Social Science Information*, 60, 188–208.

McLean, J. (2020). *Changing Digital Geographies: Technologies, Environments and People*. Cham: Springer International Publishing.

McLean, J., Maalsen, S. & Lake, L. (2022). Digital (un)sustainability at an urban university in Sydney, Australia. *Cities*, 127, 103746.

Morozov, E. (n.d). The Crypto Syllabus - About us, September 24, 2023, available at https://the-crypto-syllabus.com/about-us/

Nesi, J., Mann, S. & Robb, M. B. (2023). *Teens and Mental Health: How Girls Really Feel about Social Media*. San Francisco, CA: Common Sense.

OECD. (2023). *Empowering Young Children in the Digital Age*. Paris: Starting Strong, OECD Publishing.

Olivier, P. & Wright, P. (2015). Digital civics: Taking a local turn. *Interactions*, 22(4), 61–63.
Patrignani, N. & Whitehouse, D. (2014). Slow tech: A quest for good, clean and fair ICT. *Journal of Information, Communication and Ethics in Society*, 12(2).: 78-92,
Robinson, L., Cotten, S., Ono, H., Quan-Haase, A., Mesch, G., Chen, W., Schulz, J., Hale, T. & Stern, M. (2015). Digital inequalities and why they matter. *Information, Communication & Society*, 18, 569–582.
Schrock, A.R. (2019). What is civic tech? Defining a practice of technical pluralism. In Cardullo, P., Di Feliciantonio, C. and Kitchin, R. (Eds.) *The Right to the Smart City*. Emerald Publishing Limited, Bingley. 125–133.
Shermann, N. (2022). Crypto giant FTX collapses into bankruptcy. *BBC News*, November 11, 2022, available at https://www.bbc.com/news/business-63601213
Van Dijck, P. & De Waal, M. (2018). *The Platform Society: Public Values in a Connective World*. Oxford: Oxford University Press.
Vetter, A. (2018). The matrix of convivial technology – assessing technologies for degrowth. *Journal of Cleaner Production, Technology and Degrowth*, 197, 1778–1786.
Werbach, K. (2018). Trust, but verify: Why the blockchain needs the law. *Berkeley Technology Law Journal*, 33(2), 487–550.
Zuboff, S. (2019). *The Age of Surveillance Capitalism: The Fight for a Human Future at the New Frontier of Power*. London: Profile Books.

11

EUROPEAN STRATEGIC AUTONOMY FOR THE TWIN TRANSITION

Ambiguities and contradictions from a spatial perspective

Luis Martin Sanchez and Margherita Gori Nocentini

The rapid convergence of several crises in recent years – ecological, health-related, and, last but not least, geopolitical – has led to a rapid realignment of political and economic agendas and a wave of massive public investments in so-called strategic sectors, which in Europe are linked to initiatives such as the European Green Deal and the New Generation EU (NGEU) and the national recovery and resilience plans of member states. These plans aim to create more socio-ecologically sustainable societies by mobilising the potential of digital technologies: in this sense, the ecological and digital twin transitions are considered complementary. However, a closer look at these two agendas reveals not only the fact that in many cases they may not necessarily be aligned and synergetic but also that there may be socio-economic and environmental conflicts underway.

This chapter aims to explore the implications of the twin transition by examining the interactions between the transition rhetoric in European policies and their possible spatial effects, with a focus on the Italian case. The argument proposed by the authors, in fact, is that by adopting a spatial perspective, it is possible to bring to light the ambiguities and contradictions that mainstream discourses on the twin transition and strategic autonomy tend to obscure. The chapter focuses in particular on the contradictions of the 'European Critical Raw Materials Act', a package of measures whose main objective is to guarantee a 'secure, diversified and sustainable' supply of so-called critical raw materials for the ecological and digital transition. The package sets the objective that at least 10% of all the critical raw materials consumed in the EU be extracted within the EU's own borders (currently the rate is around 3%), thus aiming to secure a strategic advantage in the current unstable geopolitical climate. In our view, a hyper-traditional strategy based re-mining such as the one proposed is at odds with the very paradigms of the transition project. Moreover, it conflicts with deeply rooted norms and policies of

DOI: 10.4324/9781003441311-13

territorial protection in the European Union (EU) and, as shown in the course of the chapter, is also in conflict with local communities and other economic sectors rooted in the territories. The aim of this chapter is therefore to discuss, through the description and analysis of one re-mining project in Italy the actual and potential ambiguities and contradictions of European policies related to the twin transition.

The chapter is organised as follows: the first section presents the rhetoric of the twin transition as it has been constructed in EU policy discourses as a response to the multiple crises faced by European member states. The following section focuses on the concept of strategic autonomy and its recent implementation through the EU's Critical Raw Materials Act, which addresses the issue of European dependence from third countries for the supply of critical resources for the ecological and digital transitions, and is in this sense part of and supporting the broader twin transition agenda. The third section presents the Italian situation with respect to critical raw materials and introduces the setting for the case study of the re-mining project in Punta Corna, Piedmont, which is then described in greater detail in the following section. The chapter concludes with a discussion on the implications of the case study, focusing especially on the spatial effects and the contradictions generated by decisions taken within the European twin transition framework.

Twin transition agendas: the EU's response to a permanent state of crisis

For the past several years, the EU has been experiencing a 'polycrisis' (Zeitlin et al. 2019), a series of crises concerning multiple policy domains which have exacerbated internal divides among member states and threatened the Union's internal cohesion. The financial and sovereign debt crises, the refugee crisis, as well as internal political challenges such as Brexit and democratic backsliding, have been recently compounded by the effects of the COVID-19 pandemic and the Russian invasion of Ukraine and the resulting geopolitical tensions and energy crisis. This has also been happening against the backdrop of worsening climate conditions and extreme weather events linked to human-generated climate change. Over the course of the last few years, many countries in Europe and across the globe have experienced record-breaking temperatures, drought, and flooding, which have once again brought to the forefront the urgency of taking action not only to prevent further global warming but also to adapt and prepare for the often devastating effects of climate change which are already occurring.

In recent years, EU political and economic agendas have realigned themselves to address these crises, and in particular have focused on catalysing post-pandemic economic recovery, accelerating the ecological transition, and ensuring strategic autonomy for Europe in certain sectors. While mobilising a variety of policy instruments and funding schemes, EU responses to these crises have been constructed as mutually reinforcing and generating co-benefits, as for instance in the case of the 'green' focus given to the investments for post-pandemic recovery in the most hard-hit European economies. The idea is that of 'building back better', using investments related to crisis response, as in the case of the COVID-19 pandemic, to advance and accelerate the transition of European societies towards more environmentally sustainable and digitally connected economies and lifestyles.

In this respect, one particularly present rhetoric is that of the 'twin transition', that is the idea according to which the ecological and digital transitions should be mutually reinforcing and managed in a proactive and integrated way (Muench et al. 2022). In this framework, new technologies are seen as key to implement the transition to a low-carbon emitting economy, for instance through new monitoring and tracking systems to provide real-time information, simulation and forecasting tools to increase our prediction capacities, and virtualisation of production and consumption processes aimed at reducing the environmental impact of economic activities (ibid.). Technological innovation is seen in this sense as a sort of panacea for the multiple and compounding crises that the EU is facing, and first and foremost the climate crisis. However, the agendas advocated for in this framework often avoid addressing the potential contradictions and conflicts that may arise between the two objectives of digitisation and ecological transition.

Within the twin transition framework, from 2019 onwards, the EU has attempted to react to the several crises that have hit by working on three main axes: the policies of the European Green Deal (2019), the Next Generation EU fund (following the COVID-19 pandemic), and various policies and funding aimed at achieving so-called strategic autonomy (Sanahuja 2022). Strategic autonomy, as we will explain below, in fact not only is limited to defence and security aspects as in the past but now also includes issues related to digitalisation, energy security, and industrial policy in general.

The European Green Deal, launched by the European Commission in December 2019, takes as a historical reference Franklin D. Roosevelt's New Deal, a large-scale public investment programme that the U.S. president implemented from 1933 to 1939 to counter the effects of the 1929 economic crisis. The term has already been in use for some time in the United States and, since the 2008 financial crisis, also in Europe, mainly by environmentalist political groups such as the Greens in the European Parliament. Like its illustrious predecessor, the European Green Deal (EGD) is proposed as an ambitious and all-encompassing strategy that, through legislative reforms and significant funding, places the transition to low-carbon and sustainable socio-economic models at the centre of European policies. However, unlike many previous sectoral initiatives, it places the fight against the climate and environmental crisis as the cornerstone of long-term European policy, with the goal of achieving climate neutrality for the Union by 2050, while respecting the international commitments undertaken within the framework of the Paris Agreement (Sanahuja 2022). In the European Council Conclusions of 12 December 2019, it was reported: 'The transition to climate neutrality will offer significant opportunities, e.g., potential for economic growth, new business models and markets, new jobs and technological development'.

Among the main initiatives of the strategy are the European Climate Law, which presents the legally binding target of climate neutrality for member states by 2050, the EU Climate Adaptation Strategy, the EU Biodiversity Strategy 2030, the Industrial Strategy for Europe, the Action Plan for the Circular Economy, the Producer-to-Consumer Strategy, the Just Transition Fund, the new regulation on batteries and battery waste, the 'Fit for 55%' package[1] which aims to translate the ambitions of the Green Deal into legislation, as well as many others.

With the arrival of the COVID-19 pandemic in early 2020, Europe was abruptly faced with another crisis. The COVID-19 pandemic and the severe health and economic consequences in many EU countries represented a watershed for many personal and collective stories, as well as for European institutions and national governments who immediately launched a series of policies to combat the health crisis. The most striking example was the Next Generation EU (NGEU) fund, approved in July 2020 by the European Council. NGEU is a wide-reaching and ambitious funding package, providing investments for over 800 billion euros to be distributed among EU member states and used to jumpstart the post-pandemic economic recovery.

NGEU perfectly exemplifies the EU's twin transition ideal, as it is based on the idea that the pandemic should be used as an opportunity to 'emerge stronger' and 'transform our economies and societies, and design a Europe that works for everyone' (EU 2023). In this sense, NGEU was conceived less as short-term financial aid for countries in distress, and instead more as a package of long-term investments to accelerate the EU's achievement of certain policy objectives, related especially to the ecological and digital transitions. In fact, 30% of NGEU funding is allocated to tackling climate change and supporting green projects, especially through carbon emission reduction, strengthening production of clean energy and the availability of sustainable transport, environmental protection and biodiversity enhancement, and promotion of sustainable food production and consumption. In the Commission's view, these objectives are also supported by technological innovation. In fact, the digital transition is NGUE's second pillar, as the restrictions to mobility and social life implemented to block the spread of the coronavirus highlighted the importance of digitisation across all areas of EU economy and society to ensure continuity in services, work and school life, as well as social connections during lockdown periods (EC 2020).

EU Strategic Autonomy and the Critical Raw Materials Act: fundamental puzzle pieces for the twin transition

Closely connected to the twin transition ideal is that of strategic autonomy for Europe. In the context of a crisis of globalisation, several factors have drastically accelerated the EU's shift towards a 'geopolitical Europe' and put the concept of strategic autonomy back at the centre of the political debate. Among these factors are the deterioration of the relationship with China during Xi Jingping's presidency – a relationship that European Commission President Ursula von der Leyen has defined as 'systemic rivalry and economic competition', but above all the Russian invasion of Ukraine in February 2022. The emergence of a war of aggression in Eastern Europe has in fact been, in the words of President von der Leyen (2022), 'a real watershed for Europe'.

The concept of strategic autonomy has been in use by EU institutions for some time, initially developed in the defence industry and long confined to security matters. The conclusions of the November 2016 European Council defined strategic autonomy as the 'ability to act autonomously, if and when necessary, and with partners, when possible'. Since then, strategic autonomy has been extended to include new issues beyond defence, including those related to economic, logistical and technological matters. This has been motivated by the crises faced by the EU in recent years, first the COVID-19 pandemic and later the Russian invasion of Ukraine. The idea of strategic autonomy is therefore also mobilised in the EU as a response to an increasingly conflictual geopolitical context and a general crisis of the globalisation paradigm, exemplified for instance by the rise in politically reactionary sentiments and a growth in nativist attitudes against foreign threats (Sparke 2022).

On a policy level, especially in the sectors of energy security, industrial policy, logistics, and technology, in recent years, there have been efforts by EU institutions to review available instruments and make them more effective in strengthening the Union's political and economic autonomy vis-a-vis other major blocs, especially China and the United States. This process was accelerated in 2020 by the COVID-19 pandemic, which immediately revealed how a health crisis can have strong geopolitical connotations and greatly influence the strategic direction taken in policymaking. As stated by Josep Borrell, High Representative of the Union for Foreign Affairs and Security Policy, in 'Why European strategic autonomy matters' (2020): 'As such, neither masks, nor reagents, nor antibiotics are strategic products. However, when produced by a very small number of countries which turn out to be potential strategic rivals, they become strategic products'. For example, following the end of the most critical phase of the pandemic and the beginning of the recovery of global demand, the supply chain crisis began and resulted in a global lack of semiconductors, which are mostly produced in Asian countries. A similar – though certainly more politically motivated – process occurred recently with Russia's instrumentalisation of gas flows to Europe after the start of the war in Ukraine and the ensuing energy crisis. Both of these crises have accelerated and reinvigorated policies related to European strategic autonomy.

One of the most important sectors affected by the new turn towards European strategic autonomy, recently brought to light by the semiconductor crisis, is that of the extraction and refining of rare metals, which are indispensable resources for the digital and ecological transition and in general 'for the EU's industrial ecosystems' (Borrell 2020). The EU is currently almost absent from the rare metal extraction industry at a global level and is highly dependent on imports from other countries, first and foremost China, which controls 90% of the world production of these raw materials (Il

Sole 24 Ore 2023a). The strengthening of strategic autonomy for Europe therefore also entails a rethinking of its supply chains in a sector that is essential for the green and digital transitions, but where the EU is currently in a position of vulnerability.

In her State of the Union address of September 2022, President von der Leyen, speaking of Europe's dependence on Russian gas, also included the risk of dependence on third countries for the supply of rare earths and announced new European legislation on strategic raw materials, the so-called Critical Raw Materials Act. This strategy aims to strengthen the European supply chain through agreements with several countries, including Canada, Japan and Vietnam,[2] to increase material recycling rates and – the most controversial part of the plan – the development of new mines within European borders and the reopening of old mining sites (re-mining) through the use of new technologies.

In March 2023, the European Commission outlined the contents of the Critical Raw Materials Act, the main objective of which is to ensure a 'secure, diversified and sustainable' supply of so-called critical raw materials, that is materials that are crucial for both the ecological and the digital transition. The Critical Raw Materials Act especially aims to reduce Europe's reliance on a single country for these types of materials, in consideration of today's complex geopolitical environment. European dependence on the import of critical materials is marked and also extremely country-specific: 98% of heavy rare earths and 99% of light rare earths, as well as 93% of magnesium, are imported from China; 68% of cobalt from the Democratic Republic of Congo (DRC),[3] 78% of lithium from Chile, 71% of phosphorus from Kazakhstan, 64% of bauxite from Guinea and 62% of antimony from Turkey (Il Sole 24 Ore 2023a). This high dependence on non-European countries is not only a problem in terms of economic vulnerability but also has strong ethical and environmental connotations: for instance, cobalt-mining activities in the DRC have repeatedly been associated by Amnesty International and other associations with child labour exploitation, violation of human rights, and devastatingly negative environmental impacts.

The European strategy aims on the one hand to increase extraction and processing capacity within Europe's own borders – to contrast its almost total dependence on imports – and on the other hand to diversify trading partners. The European way to 'critical raw materials' is also characterised by a commitment to constructing a supply chain which minimises negative externalities on the environment and focuses on circularity. There are three main targets set by the Critical Raw Materials Act: the first is that by 2030 at least 10% of the critical raw materials consumed in the EU must be extracted from European mines, while currently the figure is 3% (Bourgery-Gonse 2023); the second objective is that at least 40% of the critical raw materials consumed in the EU will have to be refined in Europe; finally, the third target is that at least 15% of the critical raw materials consumed in the EU will have to come from recovery and recycling activities (Circular Economy Network 2023). The Critical Raw Materials Act also provides an updated list of raw materials considered critical, totalling 34.[4] Sixteen of these are considered 'strategic' and they include, for example, copper, cobalt, lithium, natural graphite, nickel and rare earths. One of the targets for 2030 is that no more than 65% of the annual consumption of each of these strategic materials should be met by a single supplier country.[5]

In the communication 'A secure and sustainable supply of critical raw materials in support of the twin transition' of the European Commission (2023a), the unprecedented growth in demand for critical raw materials needed for the twin transition is illustrated with some projections. According to EU estimates, the demand for rare earths is expected to increase 4.5-fold by 2030 and 5.5-fold by 2050, while the batteries that power electric vehicles are expected to increase demand for lithium 11-fold by 2030 and 17-fold by 2050. As in other cases, the new EU framework for critical raw materials provisioning therefore exemplifies how the ecological transition, presented as the new clean and sustainable socio-economic development model, is in reality based on extraction processes which, as we will see in more detail in the following paragraphs, have tangible and impactful effects on local communities in Europe, their economies, and lifestyles.

Finally, as any decision-making process, these policies are permeated with political meanings which need to be brought to light and explored if we are to comprehensively assess the sustainability of our choices regarding industrial policy for the transition to a low-carbon economy. This is of great importance especially considering the fact that, within the Critical Raw Materials Act, there is a stated focus on the social sustainability aspects of critical material extraction and supply chains. In fact, the European Commission (2023b) has stated how

> improved security and affordability of critical raw materials supplies must go hand in hand with increased efforts to mitigate any adverse impacts, both within the EU and in third countries with respect to labour rights, human rights and environmental protection.

According to the proposed regulation, the sustainability of projects implemented within the Critical Raw Materials Act framework should especially be assessed with regard to

> minimisation of environmental impacts, the use of socially responsible practices including respect of human and labour rights, quality jobs potential and meaningful engagement with local communities and relevant social partners, and the use of transparent business practices with adequate compliance policies to prevent and minimise risks of adverse impacts on the proper functioning of public administration, including corruption and bribery.
>
> *(European Commission 2023c)*

The case study presented below aims to catalyse a discussion on these sustainability aspects, both from an environmental and social point of view, and to analyse the local effects of EU-level policies for critical raw materials extraction.

Re-mining Italy: a return to extended extractivism?

Italy, despite having a negligible mineral production, is rich in mineral resources. According to the Ministry of Enterprises and Made in Italy, Italy has availability of 16 critical raw materials out of the 34 defined by the European Commission's Critical Raw Materials Act. However, Italy's metal mines – all of which operate underground – underwent a gradual process of decommissioning in the 1970s for various reasons, including the progressive increase in labour costs compared to developing countries, the development of deep mining sites, and the energy crisis which made production costs uncompetitive in Italy (Madeddu 2020).[6]

The strategy proposed by the Ministry of Enterprise and Made in Italy, together with the Ministry of the Environment and in line with the European strategy, aims to rehabilitate old mines and open new ones when necessary, with the overarching objective of responding to geopolitical and economic trends related to the increase in demand for these materials in the coming years, especially to support the ecological and digital transitions. At the time of writing, the two ministries have already started mapping potential extraction sites, including now inactive mines, and aim to complete this phase by the end of 2023. According to the ministries, however, before any mining activity can begin, it is also necessary to invest in recovering the necessary technological capacity, which is currently not present in Italy (Il Sole 24 Ore 2023b).

Rare earth deposits in Italy are found mainly in the Alpine regions, from Friuli to Piedmont, and in Liguria, Tuscany, northern Lazio, Abruzzo and Sardinia. Cobalt can be found in Friuli, and magnesium and copper in Veneto. In Trentino, cobalt, manganese, magnesium, barytes, and copper can be found, while Lombardy has copper, barytes, cobalt, and beryllium. Under the Piedmontese Alps, there are cobalt, graphite, and manganese. Liguria has the largest Italian deposit of titanium, in the Beigua Regional Natural Park, one of UNESCO's European Geoparks, as well as copper, graphite, manganese, and baryte. Tuscany is rich in copper and antimony, and also has manganese, magnesium, and lithium. In northern Lazio, there are some deposits of lithium, cobalt, manganese, and baryte. Baryte is also present in Sardinia, along with copper and antimony. In the Abruzzo Apennines there are several bauxite deposits and one manganese deposit. Bauxite is also found in northern Campania and in various areas of Apulia. In Calabria, there are manganese, baryte, and graphite, in Sicily, antimony and manganese (Il Sole 24 Ore 2023b).

The possibility of reopening these mines in Italy after years of inactivity brings up a number of questions and issues concerning the sustainability of these projects and their fit with the views and trajectories of the local communities.

For instance, reopening mines can be seen as at odds with the profound repositioning that the Italian economy has implemented over the last 30 years towards what Boltanski and Esquerre (2019) call the 'enrichment economy'. According to the authors (2019), this general economic repositioning of Western European countries, including Italy, towards niche markets and in general towards the high end of the supply range (the so-called luxury market) represents a break with the Fordist development model linked to mass production that had characterised the decades immediately following World War II. In contrast to mass production, which legitimised itself in democratic terms, the economy of enrichment aims to exploit the purchasing power of those who can access luxury goods. In the enrichment economy, goods are valued not so much according to their usefulness or robustness, as in the case of industrial products, but for their uniqueness. This uniqueness is in most cases associated with territorial elements, which guarantee their being authentic, traditional, and singular. In an economic model of this kind, the conservation, valorisation, and 'patrimonialisation' of territories therefore play a fundamental role.

It is no coincidence that 'patrimonialisation' has become in recent years in Western countries a proper 'territorial development' strategy, aiming to reveal and exploit previously ignored or undervalued territorial assets and enhance their potential. This type of strategy has in many cases transformed 'dormant' heritage into active heritage, stimulating the ability of actors to 'appropriate history, even at the cost of transforming it' (Boltanski and Esquerre 2019). Thus, heritage, history, tradition, identity, past, and memory are stripped of their complexity and contested nature and become devices for enhancing the value of goods and services in order to 'specify' and 'differentiate products and services from their competitors' (*Ibid.*).

The irreconcilable contradictions between this new economic model – that is, that of the Made in Italy brands put forwards especially by small and medium-sized enterprises – and a possible return to an extractivist paradigm, which would have a great impact on local communities, are therefore evident. As pointed out by Pank (2022), in fact, 'the production of operational landscapes for (…) mining requires the radical rearrangement of materiality, property relations, value, and people'. The following paragraphs illustrate these possible consequences in the case of the re-mining project in Punta Corna.

A contested future for Punta Corna: neo-extractivism, depopulation, and territorial heritage conservation

Punta Corna, a peak in the Piedmontese Graian Alps with an altitude of 2960 metres, is in some ways a typical Italian mountain area. The peak is located in the Valli di Lanzo, between the Viù Valley and the Val d'Ala. These valleys, unlike the neighbouring Val di Susa where the Agnelli family founded Sestriere in the 1930s, kick-starting mass winter tourism in Italy, have not experienced comparable tourist development. However, their fragile economy is linked, albeit to a lesser extent, to the tourism sector, especially slow and nature-related tourism, with a high number of second homes in the area. The sheep-farming sector, as well as the economy of the *malghe*, small high mountain farms for summer grazing and cheese production are also well-rooted.

In this seemingly pastoral landscape, in 2019 a licence to explore for cobalt, nickel, copper, silver and associated minerals on the Punta Corna site was granted by the Piedmont Region to the Australian mining company Altamin. Altamin is a 'junior miner', a mining company that carries out a few targeted projects, which are also related to research and surveying activities to identify possible mining locations. In Italy, Altamin has two subsidiaries, Strategic Minerals srl and Energia Minerals srl, and is active in four research projects. The most advanced project involving the excavation of zinc at Gorno, in the Province of Bergamo, was rejected in June 2022 by the Ministry for Ecological Transition (Pons 2022). At Punta Corna, the licence was instead renewed in December 2022.

The mining history of Punta Corna is not recent. From 1753 until the 1930s, it was the largest cobalt mine in Europe and most of the settlements in the area were founded by miners (Forti 2023). At the time, cobalt was used as a colouring pigment, but with the emergence of synthetic dyes in the early twentieth century, the mine was decommissioned. Since then, cobalt's uses have diversified, and it has become an indispensable material in numerous industrial processes due to its specific physical and chemical properties. For example, it is an essential component in super alloys used in the production of jet engines, gas turbines, helicopter blades, and power plant turbines; it is also used to produce high-strength permanent magnets used in loudspeakers, electric motors, and other applications; alloys containing cobalt are used in the production of high-strength cutting tools, such as drill bits, cutters, and blades. Cobalt is also used in the production of adipic acid employed in nylon synthesis and other oxidation reactions.

Above all, cobalt production has increased significantly in recent decades due to the growth in demand for lithium batteries for portable electronic devices, electric vehicles, and other applications. It is also expected that the growth of electric vehicle production, with the ban on the sale of new petrol and diesel cars in the EU from 2035,[7] will stimulate a strong demand for lithium batteries, and therefore for cobalt as a key component of many rechargeable batteries. It is therefore no coincidence that in 2020 the EU included cobalt among the raw materials of 'strategic interest', together with lithium, coltan, and rare earths, specifically in support of the energy and digital transition.

The Punta Corna mine is located between two municipalities in the Lanzo Valleys, Usseglio, and Balme. The municipal administrations have taken different, and in some ways opposing, positions with respect to the possible extractive development of the area. These two positions are exemplary of the development strategies of mountain areas in Italy today: on the one hand, a traditional model, a legacy of the Modern Age, linked to extractive activities and energy production; on the other hand, a model related to Boltanski and Esquerre's (2019) 'enrichment economy', which is instead very much associated with niche markets and processes of territorial patrimonialisation and valorisation.

In the ostensibly pastoral territory of the municipality of Usseglio, modernity[8] has left its mark with the construction in the early 1930s of a large dam for the production of electricity – currently managed by Enel electric company – and the creation of the artificial lake of Malciaussia, which covered the old hamlet of the same name. The construction of the dam brought infrastructural development to the Lanzo Valleys, with the building of roads, railways, and the creation of numerous jobs in the secondary and tertiary sectors.[9] The administration of Usseglio has expressed hope that the development of the Punta Corna cobalt mine will have a similar effect on its territory (Forti 2023), which like many mountainous areas in Italy today is suffering from great infrastructural deficiencies – especially concerning digital infrastructure[10] – and is subject to widespread depopulation.[11] The local administration in fact does not see the potential development of the new mine as a threat to the economic activities rooted in the area – especially those related to mountain tourism – but as a stimulus to the local economic system (*Ibid.*). However, the contradictions between a production model linked to mining and another associated with the tourism sector are evident. The risks of impoverishment of environmental quality in territories reduced to mining machines, which has been extensively researched,[12] are difficult to reconcile with tourism.

The other municipality affected by the mining project, located on the other side of Punta Corna in the Val d'Ala, is Balme. Here, the position of the local administration is opposite to that of Usseglio and already in June 2020 the municipal council passed a resolution opposing Altamin's mining exploration (Pons 2022). The local economic sector is built around niche nature tourism and land valorisation, and in fact one area in the municipality has been part of the 'Pian della Mussa' Natura 2000 Network nature protection site since 2016. However, Balme also already has another type of extractivist activity within its boundaries: water bottling. The production of the small company Pian della Mussa Srl, which bottles spring water, and its relationship with the surrounding territory, which is of great environmental value and has a strong vocation for tourism, is part of that enrichment economy (Boltanski and Esquerre 2019), an economic model which has much to do with the processes of valorisation and patrimonialisation of the territory and that is therefore naturally conflicting with an extractive economic activity such as mining.

However, these processes of 'territorial patrimonialisation' are not risk-free. In fact, the identification of a place as an asset implies its 'valorisation' (Lazzarotti 2003), which also brings about the risk of commodification of territorial identity and the exclusion of anything that is outside a certain well-defined narrative. If, on the one hand, it is certainly the case that these valorisation processes may reach their culmination with the recognition by international organisations (the EU's Natura 2000 Network in the case of Pian della Mussa), at the same time, this labelling can lead to an 'iconic reduction' (D'Eramo 2017) of the territory itself, with its identity and image, and thus to its homologation in the name of greater tourist accessibility (*Ibid.*). On the other hand, the creation and implementation of 'protection' protocols, aimed at the conservation and environmental protection of these areas, may also result in closing off any possibility of shared modification and transformation of the territory, even by local actors themselves (Vassallo and Martin Sanchez 2022).

In Balme, therefore, two antithetical economic models and two different visions of the territory come into irremediable conflict. Here, the mountain is no longer seen as a productive space, as it had been in the years of modernisation, but as a space to be safeguarded, even at the risk of reducing the territory to an image, with no possibility of change and evolution. At the same time, the envisioned response to the risk of depopulation and territorial abandonment is strongly linked to the strengthening of infrastructure, first and foremost digital, which could attract new urban populations and boost the local tourist system.

What is evident from this brief case study description is that there is no unitary vision for the future of Punta Corna. On the one hand, especially in the case of Usseglio, re-mining is not considered in contrast with current economic activities and instead is seen positively as there is the hope that new industrial activities will attract further investments in the area. On the other hand, the position expressed by Balme municipality has a much stronger conservationist streak and is more rooted in an 'enrichment economy' model. Across the board, however, there is the perception that strengthening the attractiveness of Punta Corna to outside residents in order to contrast depopulation is necessary, whether it be through new economic opportunities brought by the mining sector or through promotion of tourism.

It seems paradoxical, however, that the hopes of certain local actors for the survival of this territory, especially in the face of depopulation and lack of infrastructure, lie precisely in greater development of digital infrastructure for which the mine that they are so clearly opposed to is necessary. Once again, we see here at work what Iapaolo, Certomà, and Martellozzo in the Introduction of this book (this volume, p. 3) have described as "the 'fantasies of dematerialization' associated with the digital economy, a notion that often obscures its reliance on a vast and finite assortment of 'materials and planetary resources'".

What's more, the different visions for Punta Corna's future expressed by local actors, as well as the contradictions between them that have been brought to light by the new mining project and briefly described here, are currently lacking a suitable space for discussion. Once again, the mining project is a case of top-down policy deriving from a European framework, the Critical Raw Materials Act, developed by the two competent national Ministries, and operationally led by a private business, based on a licence awarded by the Region. Local actors are instead excluded from the process and are stripped of any real decision-making power on the matter, as municipalities only have a consulting capacity (Forti 2023).

Twin transition territories: ambiguities and contradictions of top-level strategies on the ground

The case described above shows how the policies developed at the EU and national levels in the framework of the twin transition, once they begin to be implemented on the ground, may come into friction with a series of factors and actors of a geopolitical, social, ecological and economic nature. Thus, the most contradictory and sometimes obscure sides of the transition project emerge, which, despite an all too entrenched mainstream narrative, is not a pacified and good-for-all project. On the contrary, the transition project, like any new paradigm, reconstitutes hierarchies, priorities, relations, and values, generating conflicts between actors with diverging interests and intentions. According to our interpretation, at least three issues emerge from this case which show the ambiguities and contradictions of the transition project in European territories.

The first is the possible contradiction between the digital and ecological transitions themselves. Identifying all the environmental consequences of digitalisation is challenging, as it involves complex processes of energy consumption, land use, labour, and material transformations across multiple sites and scales (McLean 2019). Still, it has been amply demonstrated that the digital industry has a significant (and growing) environmental and social impact (Pearce 2018). We believe that this case study supports the idea that, in order to truly advocate for sustainability, the twin transition rhetoric should significantly engage with the environmental impacts of digitalisation itself.

The second contradiction revealed by the case study is between different (local) visions for the future of Punta Corna, which subscribe to different views concerning development models. On one hand, one model is based on (over)protection of natural areas and development of tourism linked to environmental quality; on the other hand, territorial development through the reactivation of industrial and extractivist activities. As we have seen above, both of these models generate further risks in themselves; however, the main point is that they are at odds with each other and diverge with respect to the consideration given to the Punta Corna re-mining project.

Finally, further contradictions and possible conflicts are fuelled by the unequal distribution of decision-making powers across the governance structure. The Italian national-level and European institutions are aligned in their objectives concerning strategic autonomy in the sector of critical raw materials. However, the institutional architecture of the jurisdictions and actors affected by the envisaged extraction processes is lacking in representation of the territorial and local levels. In the words of Usseglio's mayor, 'Let's be clear: the municipal administration has no say in the matter. It is the Region that grants mining licences, and it is the Ministry of the Environment that authorises them' (Forti 2023). Once again, however illuminated in their aims of achieving a low-carbon emitting economy in Europe, these policies are fundamentally top-down in nature and based on partnerships between public actors located at the top of the governmental hierarchy and large private actors, such as businesses, with the exclusion of local communities and administrations. We consider this particularly problematic as one of the foci of the Critical Raw Materials Act concerns the social sustainability of supply chains, which also includes meaningful engagement with local communities.

The twin transition model fits into the ecological modernisation rhetoric (Bianchetti 2021), as the transition project is constructed as fundamentally managerial, technocratic, and apolitical. As pointed out by Swyngedouw and Heynen (2003), the ecological transition paradigm is based on faith in scientific solutionism and contributes to the depoliticisation of environmental demands. In this sense, the twin transition ideal is based exactly on this belief, that is, that technical fixes can be used to comprehensively address the problems related to human-generated climate change. This idea instead does not allow for a political perspective to enter the equation, such as considering the power relations that underlie and shape processes of environmental degradation and, for instance, determine an unequal distribution of the associated costs.

According to the authors of this chapter, not only does the twin transition rhetoric further reinforce this depoliticising trend but also does the pursuit of strategic autonomy for Europe in the face of an increasingly tense geopolitical environment. Strategic autonomy objectives now include a sufficient supply of rare raw materials, as well as other critical resources, as the very foundation for Europe's continued development, for instance in support of the digital transition. While on one hand this can be seen as a way to avoid procurement of these materials from socially and environmentally unsustainable supply chains, for instance due to gross violations of human rights, on the other hand, as we have seen, these policies are not immune to generating contradictions and conflicts in the areas where they are and will be implemented.

These contradictions, however, are currently obscured in the general narrative, and bringing them to the forefront is of fundamental importance to evaluate and determine the real sustainability of policies aimed at achieving the ecological transition, by taking into account the environmental and social impacts on the local communities, which will be affected by new extractivist activities. In this sense, having a spatial perspective as well as a regard for the procedural aspects of how decisions in this realm are taken can help mitigate the risk of overlooking and silencing the perspectives, views, and desires of those people and places that are directly bearing the consequences. Failing to take into account the spatiality of top-level policies, as well as the specific conditions of the places where they will be implemented, does not allow for the social sustainability of their objectives and results to be adequately taken into account. The current state of permanent crisis (Agamben 2020; Bauman and Bordoni 2014; Latour 2020; Pulcini 2013; Tsing 2015) especially obliges us to critically rethink the territorial effects of the twin transition project (Décleve et al. 2020), which once again is at risk of perpetuating a technocratic project that renounces any radical and transformative vision and instead, by mobilising digital technologies reproduces mechanisms of exploitation and exclusion that are simply repackaged as green, sustainable, safe, healthy, and resilient.

Notes

1 The package, proposed by the European Commission on 14 July 2021, consists of a series of proposals to revise climate, energy, and transport legislation and to implement new legislative initiatives to align EU regulations with its new climate goals. The name

refers to the interim target of cutting carbon emissions in the EU by 55% (compared to 1990 emission levels) by 2030.

2 As Certomà, Iapaolo, and Martellozzo state in their introduction to this book,

> The race for rare and precious metals pushes the frontiers of extraction further in liminal geographical areas where people work in miserable environmental conditions with limited or no control of the safety standards, especially in the global South, to feed the competition for global connectivity.

3 Almost three quarters of the world's cobalt production takes place in the Democratic Republic of Congo. Most mining operations are controlled by large multinationals, mainly Chinese, but at least 20% of the exported material comes from 'artisanal' mines that use child labour and do not respect fundamental labour and human rights conditions.

4 In the European Commission's draft regulation, 34 'critical raw materials' are listed: antimony, arsenic, bauxite, baryllium, bismuth, boron, cobalt, coking coals, copper, feldspar, fluorspar, gallium, germanium, hafnium, helium, heavy rare earth elements, light rare earth elements, lithium, magnesium, manganese, natural graphite, nickel-battery grade, niobium, phosphorite, phosphorus, platinum group metals, scandium, silicon metal, strontium, tantalum, titanium metal, tungsten, and vanadium. Of these 'critical' raw materials, 16 are considered 'strategic': bismuth, boron-metallurgical grade, cobalt, copper, gallium, germanium, lithium-battery grade, magnesium metal, manganese-battery grade, natural graphite-battery grade, nickel-battery grade, platinum group metals, rare earth elements for magnets (Nd, Pr, Tb, Dy, Gd, Sm, and Ce), silicon metal, titanium metal, and tungsten.

5 Other measures envisaged in the framework of the Critical Raw Materials Act are: the simplification and acceleration of regulations for extraction projects: a maximum of 24 months for extraction permits and 12 months for processing and recycling permits; simplified procedures supported by easier access to funding lines; greater coordination between Member States in order to have an accurate picture of national reserves; investment in research and training with the creation of a Raw Materials Academy to develop the new skills needed by workers engaged in this sector. In addition, measures are planned to increase recycling rates and the use of secondary raw materials.

6 In 1982, the Italian state (unsuccessfully) attempted a revival of the sector by deploying substantial public resources to support activities from mining research to production (Madeddu 2020).

7 On 14 February 2023, the European Parliament gave final approval to the measure banning the sale of vehicles with combustion engines from 2035. The measure is part of the 'Fit for 55' package to halve pollutant emissions in the EU by 2030 and sets a target of zero emissions from new cars and vans for sale in the EU from 2035.

8 In modernity, mountain territories, and in particular alpine areas, have been extraordinary places of experimentation and utopia in the fields of architecture and urban planning: from Viollet-le-Duc's projects, to those of Portaluppi, to the expressionism of Taut and Hablik, to the modern projects of Figini and Pollini, to the 1960s landscape design of Halprin.

9 See A. De Rossi. *The construction of the Alps. The twentieth century and Alpine modernism (1917–2017).*

10 On the state of digital infrastructures in the Italian Alpine territories with a specific focus on Piedmont, see F. Cavallaro & A. Dianin, Beyond the transport infrastructure: the paradigm of accessibility.

11 In Usseglio, there are 200 residents (as of 31 December 2019), with an advanced average age. On the depopulation issues of Italy's inner areas, see A. De Rossi ed., *Riabitare l'Italia. Le aree interne tra abbandoni e riconquiste*; Italian National Network of Young Researchers for Inner Areas Committee, *Inner Areas in Italy. A testbed for analysing, managing and designing marginal territories.*

12 See M. Arboleda, *Planetary Mine. Territories of Extraction Under Late Capitalism*; M. Bjornerud, *Timefulness How thinking like a geologist can help save the world*; N. Brenner & N. Katsikis, Operational landscapes: hinterlands of the Capitalocene; J. Sordi, L. Valenzuela & F. Vera, *The Camp and the City: Territories of Extraction.*

References

Agamben G. (2022). *When the House Burns Down*. London: Seagull Books.

Arboleda M. (2020). *Planetary Mine. Territories of Extraction Under Late Capitalism*. London: Verso.

Bauman Z. & Bordoni C. (2014). *State of Crisis*. Cambridge: Polity.

Bianchetti C. (2021). Urbanistica e Sostenibilità. In N. Martinelli & M.V. Mininni (Eds.), *Città Sostenibilità Resilienza. L'urbanistica italiana di fronte all'Agenda 2030* (pp. 27–33). Roma: Donzelli Editore.

Bjornerud M. (2018). *Timefulness How Thinking Like a Geologist Can Help Save the World*. Princeton: Princeton University Press.

Boltanski L. & Esquerre A. (2019). *Enrichment: A Critique of Commodities*. Cambridge: Polity Press.

Borrell J. (2020). Why European strategic autonomy matters. Available at: https://www.eeas.europa.eu/eeas/why-european-strategic-autonomy-matters_en (Accessed 31 July 2023).

Bourgery-Gonse T. (2023). EU unveils Critical Raw Materials Act, aiming to lessen dependence on China.' *Euractiv* (16 March). Available at: https://www.euractiv.com/section/economy-jobs/news/eu-unveils-critical-raw-materials-act-aiming-to-lessen-dependence-on-china/ (Accessed 31 July 2023).

Brenner N. & Katsikis N. (2020). Operational landscapes: Hinterlands of the Capitalocene. *AD / Architectural Design*, 90, 22–31.

Cavallaro F. & Dianin A. (2022). Beyond the transport infrastructure: the paradigm of accessibility. In C. Boano & C. Bianchetti (Eds.), *Lifelines: Politics, Ethics, and the Affective Economy of Inhabiting*. Berlin: Jovis, 88–105.

Circular Economy Network. (2023). Critical Raw Materials: The EU Strategy for Critical Raw Materials. Available: https://circulareconomynetwork.it/2023/03/17/critical-raw-materials-act/ (Accessed 31 July 2023).

Consigliere S. (2019). Archeologia della dissociazione. In P. Bartolini & S. Consigliere (Eds.), *Strumenti di Cattura. Per una Critica dell'immaginario tecno-capitalista*. Milano: Jaka Book.

Décleve et al. (2020). *Dessiner la transition*. Geneve: Métis Presses.

D'Eramo M. (2021). *The World in a Selfie. An Inquiry into the Tourist Age*. London: Verso Books.

De Rossi A. (2016). *The Construction of the Alps. The Twentieth Century and Alpine Modernism (1917–2017)*. Roma: Donzelli.

De Rossi A. ed. (2018). *Riabitare l'Italia. Le aree interne tra abbandoni e riconquiste*. Roma: Donzelli.

European Commission (2020). Europe's moment: Repair and Prepare for the Next Generation. Available at: https://eur-lex.europa.eu/legal-content/EN/TXT/PDF/?uri=CELEX:52020DC0456 (Accessed 9 August 2023).

European Commission. (2023a). A secure and sustainable supply of critical raw materials in support of the twin transition. Available at: https://eur-lex.europa.eu/legal-content/EN/TXT/?uri=CELEX%3A52023DC0165 (Accessed 31 July 2023).

European Commission. (2023b). Critical Raw Materials: Ensuring secure and sustainable supply chains for EU's green and digital future. Available at: https://ec.europa.eu/commission/presscorner/detail/en/ip_23_1661 (Accessed 9 August 2023).
European Commission. (2023c). Regulation of the European Parliament and of the Council establishing a framework for ensuring a secure and sustainable supply of critical raw materials and amending Regulations (EU) 168/2013, (EU) 2018/858, 2018/1724 and (EU) 2019/1020. Available at: https://eur-lex.europa.eu/resource.html?uri=cellar:903d35cc-c4a2-11ed-a05c-01aa75ed71a1.0001.02/DOC_1&format=PDF (Accessed 9 August 2023).
European Union. (2023). NextGeneration EU: Make it real. Available: https://next-generation-eu.europa.eu/index_en (Accessed 9 August 2023).
Forti M. (2023). La corsa al cobalto arriva in Piemonte. *Internazionale* (March 6). Available at: https://www.internazionale.it/essenziale/notizie/marina-forti/2023/03/06/piemonte-miniere-cobalto (Accessed 31 July 31).
Huitema J. (2023). EU ban on the sale of new petrol and diesel cars from 2035 explained. Available at: https://www.europarl.europa.eu/news/it/headlines/economy/20221019STO44572/il-divieto-di-vendita-per-le-nuove-auto-a-benzina-e-diesel-nell-ue-dal-2035?at_campaign=20234-Green&at_medium=Google_Ads&at_platform=Search&at_creation=DSA&at_goal=TR_G&at_audience=&at_topic=Emissions&gclid=Cj0KCQjw2qKmBhCfARIsAFy8buIS80ObdQgceI32McSaIciFBs2GH9VYqMk1QXkFUvctoCH26VmcqcIaAlvREALw_wcB (Accessed 31 July 2023).
Il Sole 24 Ore. (2023a). Terre rare e tecnologie, la rincorsa europea. *Il Sole 24 Ore* (8 March 8). Available at: https://lab24.ilsole24ore.com/terre-rare-europa/?refresh_ce=1 (Accessed 31 July 2023).
Il Sole 24 Ore. (2023b). Terre rare, ecco dove sono in Italia e cosa vuole fare il governo. *Il Sole 24 Ore*, (14 July). Available at: https://www.ilsole24ore.com/art/terre-rare-ecco-dove-sono-italia-e-cosa-vuole-fare-governo-AFsNdQD (Accessed 31 July 2023).
Italian National Network of Young Researchers for Inner Areas Committee. (2021). *Inner Areas in Italy. A Testbed for Analysing, Managing and Designing Marginal Territories*. Trento: ListLab.
Latour B. (2020). *La sfida di Gaia. Il nuovo regime climatico*. Sesto San Giovanni: Meltemi.
Lazzarotti O. (2003). Tourisme et patrimoine: ad augusta per angustia. *Annales de géographie*, 629, 91–110.
Madeddu D. (2020). L'Italia delle miniere "dimenticate": dove sono e i tesori che si estraggono. *Il Sole 24 Ore* (29 November). Available at: https://www.ilsole24ore.com/art/l-italia-miniere-dimenticate-dove-sono-e-tesori-che-si-estraggono-ADaHIB5 (Accessed 31 July 2023).
Martinelli N. & Mininni M. (Eds.) (2019). *Città Sostenibilità Resilienza*. Roma: Donzelli Editore.
McLean J. (2019). For a greener future, we must accept there is nothing inherently sustainable about going digital. *The Conversation* (16 December). Available at: https://theconversation.com/for-a-greener-future-we-must-accept-theres-nothing-inherently-sustainable-about-going-digital-128125 (Accessed 31 July 2023).
Muench S., Stoermer E., Jensen K., Asikainen T., Salvi M. and Scapolo F. (2022). *Towards a Green & Digital Future. Key Requirements for Successful Twin Transitions in the European Union*. Luxembourg: Publications Office of the European Union.
Pank F. (2022). Extended extractivism: The production of an operational landscape in Eastern Germany. *Emptiness Field Reports* (24 October). Available at: https://emptiness.

eu/field-reports/extended-extractivism-the-production-of-an-operational-landscape-in-eastern-germany/ (Accessed 31 July 2023).

Pearce F. (2018). Energy hogs: Can world's huge data centers be made more efficient? *YaleEnvironment 360* (3 April). Available at: https://e360.yale.edu/features/energy-hogs-can-huge-data-centers- be-made-more-efficient (Accessed 31 July 2023).

Pons L. (2022). In Piemonte potrebbe nascere una miniera di cobalto: la sua storia, raccontata da chi abita lì. *Rolling Stone* (3 July). Available at: https://www.rollingstone.it/politica/in-piemonte-potrebbe-nascere-una-miniera-di-cobalto-la-sua-storia-raccontata-da-chi-abita-li/648784/ (Accessed 31 July).

Pulcini E. (2013). *Caring of the world: Fear, Responsibility and Justice in the Global Age*. Berlin: Springer.

Rittel H.W. & Webber M.M. (1973). Dilemmas in a general theory of planning. *Policy sciences*, 4(2), 155–169.

Sanahuja J.A. (2022). El Pacto Verde, NextGenerationEU y la nueva Europa geopolítica. *Documentos de trabajo*, 63(2a). https://doi.org/10.33960/issn-e.1885-9119.DT63

Sordi J., Valenzuela L. & Vera F. (2017). *The Camp and the City: Territories of Extraction*. Trento: ListLab.

Sparke M. (2022). The crisis of globalisation. In R. Ballard & C. Barnett (Eds.), *The Routledge Handbook of Social Change*. London: Routledge.

Tsing A.L. (2015). *The Mushroom at the End of the World: On the Possibility of Life in Capitalist Ruins*. Princeton: Princeton University Press.

Vassallo I. & Martin Sanchez L. (2022). The ambivalent nature of productive lifelines: Values without waste and landscape without inhabitation. In C. Boano & C. Bianchetti (Eds.), *Lifelines: Politics, Ethics, and the Affective Economy of Inhabiting*. Berlin: Jovis, 240–255.

von der Leyern U. (2022). State of the Union: A Union That Stands Strong Together. Available at: https://ec.europa.eu/commission/presscorner/detail/ov/speech_22_5493 (Accessed 31 July 2023).

Zeitlin J., Nicoli F. & Laffan B. (2019). Introduction: The European Union beyond the polycrisis? Integration and politicization in an age of shifting cleavages. *Journal of European Public Policy*, 26, 963–976.

12

EXCAVATING DIGITAL (UN)SUSTAINABILITIES

Jessica McLean

Digital solutions are increasingly sought for intractable social and environmental problems, by individuals, communities and organisations, but there are a range of consequences of such inclinations that may be overlooked or not considered when turning to digital infrastructures. Amongst these consequences is deepening reliance on unsustainable practices that are often used to create and maintain digital infrastructures, as well as in the generation and storage of digital data. Addressing urgent environmental problems by turning to the digital is appealing for many reasons, including responding to a lack of decisive and effective action on a global scale to phase out fossil fuels to mitigate climate change (Carrington, 2023) and the devastating realities of living within highly stressed global environmental systems. It makes sense when facing intransigent environmental crises that individuals, organisations and communities are seeking digital interventions to (re)mediate these.

However, as the editors in the introduction and the authors of this collection's chapters outline, there are a range of contradictions, possibilities and constraints in terms of using digital technologies to try to achieve sustainability goals. The authors herein excavate a broad range of issues that are characterised as digital (un) sustainabilities and query claims of better futures via the digital in different ways. This conclusion offers a brief overview of what I perceive as the main themes put forward in this collection before offering a coda in the form of a walk through and along-with a digital thing's imagined journey. The coda traces an imagined path of a nature selfie to trace some of its 'loose ends and missing links' (Massey, 2005: 12) and delineates some of the tensions, contradictions and possibilities associated with seeking better digital and environmental futures.

The three main themes that struck me when reading the collected chapters were the gaps and affordances relating to digital participation; problematisation of the

DOI: 10.4324/9781003441311-14

forms and structures that undergird current digital technologies that aim to increase sustainability, and unpacking a range of claims that are associated with digital solutionism; and the scale and geographies of digital (un)sustainabilities. I have referred to excavating digital (un)sustainabilities in this title as all chapters aim to reveal qualities of digital infrastructures that are often hidden by discourse or rhetoric that are difficult to understand at first glance. The 'digging up' process relies upon thinking with and talking through inconsistencies and paradoxical relations that are not always obvious but are essential to critically contributing to building genuinely sustainable digital infrastructures.

Gaps and affordances relating to digital participation

With a focus on digital participatory tools, Caitlin Hafferty, Jiri Panek and Ian Babelon problematise digital democracy processes and the assumptions that underpin its practices. Despite supposed benefits of expanding participation and extending democratic processes, there are important questions about the effectiveness of digital technologies for actually achieving that end, and possible unintended consequences of such efforts. They focus on socio economic factors and contextual issues that serve to impede the attainment of comprehensive digital democracy goals. A pragmatic approach is warranted, Hafferty et al. suggest, that does not pursue a 'digital-by-default' set of practices but is more flexible and context specific: they argue that moving from proposing the digital as an all-inclusive strategy should include possibilities for hybrid and other flexible strategies to increase democratic processes.

Taking a more techno-optimist approach, Igor Calzada looks at the multiple challenges surrounding data sustainability and argues that techno-capitalist organisations continue to produce systems that are extractive and inequitable. As alternatives, Calzada suggests that Web3 is the way forward, including decentralised autonomous organisations, data cooperatives and blockchain-based decentralised data architectures as providing avenues to empower individuals and communities to reclaim control over their data. One possible tension in Calzada's argument that greater participation and ownership of digital technologies by citizens is that we do not have sufficient evidence that this will substantially increase the sustainability of these systems. They may decentre big tech from controlling the machines, but communities and individuals are also likely to make environmentally unsustainable choices given evidence from other sectors and practices. Calzada's enthusiasm for a path towards a digital utopia 'where people reclaim their data destiny and forge a more sustainable and just digital world' sounds appealing and offers a hopeful stance on digital futures.

Similarly, Mark Dyer, Shaoqun Wu and Min-Hsien Weng also focus on community participation in digital practices but this time in the context of rebuilding a city following a disaster: the Christchurch earthquakes and the infrastructural damage it induced. They argue that the process of handing over control of the rebuilding of the city to the Canterbury Earthquake Recovery Authority was flawed and grounded in assumptions that bureaucracies could deliver effective planning without grassroots citizen involvement. Dyer et al propose an alternative approach to consultation which doesn't involve that particular form of institutionalised and centralised planning but draws on artificial intelligence (AI) using large amounts of public data that could be processed to communicate forms of what they call 'shared narratives'. Similarly to Calzada, Dyer et al. view digital technologies as a solution that could increase effectiveness of public processes by circumventing entrenched structural and political problems. Interestingly, bias in terms of gathering representative data is identified as an issue in Dyer's chapter but the question of bias and limitations of AI in terms of how algorithms work, and the risks of these digital tools (Del Casino et al., 2020; Maalsen, 2023) is not evaluated.

The turn to digital technologies as solutions for environmental crises, in particular climate crises, is put under the spotlight in Machen's chapter on digital fractures and the reliance on climate models. Digital policy tools are now ubiquitous for climate responses, Ruth Machen argues, and the ways in which these are used and their actual outcomes must be critically evaluated. For example, an important pitfall in turning to the digital to achieve sustainability is that qualitative sustainability concerns – and those beyond climate issues – may not come into the picture if quantitative modelling is always the mode of measuring and assessing environmental dilemmas. These prescient concerns relating to the digital's capacity to capture qualitative information when generating models for climate management also could be extended to other spheres.

Delineating forms of and challenges to digital sustainability and tensions within these gestures

The second major theme in the chapters involves critiquing rhetorical gestures of sustainability that do not fit with actual practices in digital infrastructural practices. Cristina Viano, Guido Boella and Claudia Schifanella describe the tensions between an increasing demand for digital technologies that are environmentally and socially sustainable by civic society groups and social economies practitioners and the complex digital skills that are associated with the alternatives to big tech offerings. Drawing on interdisciplinary research involving computer science and the social sciences, Viano et al. examine the ongoing work of the Digital Territories and Communities research group on digital tools such as a civic social network and a civic blockchain wallet. They presciently point out that the increasing digitalisation of everyday life brings opportunities and challenges for greater social justice and equity. Overall, they plot pathways to steer digital platforms from individual interests and towards public goods. They are ultimately applying a careful and powerful approach to academic research that amplifies and connects existing ethical community economy projects in a performative sense. Helpfully, Viano et al. articulate the differences between environmental sustainability and social sustainability, highlighting how we are a long way from achieving 'green' digital technologies.

In a big picture chapter on digital sustainability claims by technological corporations, including the tech giants (Apple, Microsoft, Amazon, Meta and Alphabet), Mel Hogan frames claims of environmental sustainability as greenwashing. Hogan's chapter offers an analysis of 'big cloud solastalgia' and states that the greenwashing at play by tech giants deliberately obfuscates the logics of colonial, capitalist and power relations that enable their massive profits. Solastalgia refers to sadness and grief caused by environmental degradation and loss; in this case, the multiple environmental systems that are pressured by the extractive regimes facilitating the digital. This particular sense of solastalgia may be intensified by the unreasonable assertions of big tech's rhetorical gestures that they produce green ICT. Hogan powerfully elucidates the differences between nostalgia and solastalgia: 'We may be nostalgic for what was or might have been, but we are solastalgic in our embodiments of the present crisis'.

The academic narratives of AI and the extent to which it enables (or constrains) environmental sustainability and its impacts on social/democratic sustainability are considered by Irene Niet, Mignon Hagemeijer, Anne Marte Gardenier and Rinie van Est. They conclude that AI research and data has many gaps with respect to the environmental impacts of these tools but that digital technologies are far from sustainable, including in relation to the 'sourcing and shipping of rare metals and the conflicts this brings in already fragile states'. Here, digital unsustainability is intrinsically related to social instability and the uncritical use of AI deepens these issues.

Linking arguments for degrowth with critiques of unnecessary digitalisation, Sy Taffel provides a compelling analysis of the fantasies associated with dematerialisation of the digital and how they enable the perpetuation of 'exponential rates of "green" economic growth on a materially finite planet'. Rather than stopping at critique, Taffel proposes that we pursue whole system transformation towards degrowth, abandoning digital capitalism (and one might suppose capitalism of all sorts). The benefits of adopting a degrowth, or post-growth, model are multiple, according to Taffel, including achieving decommodification and greater conviviality. Digital degrowth involves reducing material and energy use in order to prevent pushing beyond planetary boundaries that are central to digital infrastructure. Specifically, Taffel argues that replacing for-profit corporate digital services with not-for-profit alternatives is crucial. Degrowth proposals are often faced with arguments that moves towards such systems just means having less overall but that it would involve having less of 'unwanted' things: for Taffel, this includes advertising, dataveillance, bitcoin, irreparable devices and more.

The scale, scope and geographies of digital (un)sustainabilities

The third theme I see emerging in this book relates to the scale, scope and geographies of digital (un)sustainabilities. Depending on the scale and geographic range that one uses to examine digital infrastructures, different pictures of possibly sustainable (or otherwise) realities appear.

In a chapter on the gaps between efforts to achieve control of digital materiality and securing environmental futures, Luis Martin Sanchez and Margherita Gori Nocentini write about the 'twin transitions' that are currently gaining traction within European political and economic agendas. The first transition involves shifting towards more digital innovations and the other aims for more socio-ecologically sustainable societies. They evaluate the decision-making accompanying the re-opening of a cobalt mine in Italy, and point out the tensions and ambiguities of what is actually going on to support expanding digital infrastructure usage and meeting requirements of legal interventions to secure the transitions, in this case the European Critical Raw Materials Act. The contradictions in the Act, according to Sanchez and Nocentini, include economic and environmental factors but I will just mention the latter here. The reopening of this cobalt mine in northern Italy, called Punta Coma, involves an exploratory licence to an Australian mining company, and is closely examined by the authors and presented as generating conflicts between local people's hopes for maintaining tourism and natural values and other scaled intentions for digital futures based on extractivist goals. Sanchez and Nocentini focus on the spatiality of these processes and scale-based differences are at the core of these contradictions and tensions.

In another chapter looking at mining practices, Stephen Cornford suggests that we need to shift our focus to the scale of the landscape when thinking about the relationships between digital technologies and environmental change. And not just the landscape as a general fact, but digging into the ways that landscapes are remade for our digital consumption practices. Cornford states that there is an imperative 'for the operative gaze to be exerted upon the collateral' and that the circular economy, if to be effective, must incorporate actioning the ever more significant volumes of tailings that are being produced as so-called by-products of mining. The physical realities of exponential Internet usage are best understood at the landscape scale, Cornford argues.

Coda: the nature of digital things

My last nature selfie was taken at dawn, on the banks of the Cooks River, Gadigal Country, Australia; I looked upwards to the golden pink clouds rather than down to the greasy urban river to frame and take it. The litter that lined the riverbed and pock-marked the water surface are invisible in that photo and so is the diffuse digital infrastructure that made it possible. In that small moment, I made what looks like a smooth and whole digital world in that partial nature selfie.

Digital worlds tend to gloss over the spikes in our infrastructure and the grime of the everyday; when digital infrastructure works, we don't want or need to know more than that our devices are here to use, and that they are easy to wrangle. But the digital infrastructures that produce these worlds are vivid and troubling in important ways that invite our attention, at least for a short while.

Come for a walk with me, then, through the substrate of the digital world that afforded my last nature selfie. It will not be a complete story of that selfie's digital world, but it shares some openings into it. That selfie only came into being because of machines talking to each other through cables and radio waves via static infrastructure that are far from neutral in their environmental impacts.

We'll go along that path my last nature photo took, follow its trajectory as it moved from my smartphone to Instagram, and onwards again, temporarily landing on someone else's device. Let's dig into one particular digital world for a spell, forged of glass and metals and plastic, to see the grit that forges the smooth surfaces therein.

The stops that we make in this digital world walk-along will be fleeting, only a partial meatspacing of the digital and a semi fleshing out of the ethereal. We will just glimpse a crescent moon slice of all possible digital worlds.

So, to start with that nature selfie, I took against the river at dawn: mediated by my smartphone but, when shared, #unfiltered. The smartphone is the most ubiquitous actor in this digital ecosystem and engineer Professor Lotfi Belkhir is our guide at this stop. He's shown that a smartphone's chip and motherboard consumes the most amount of energy of all components of digital ecosystems because of the rare and precious metals that make them (Belkhir and Elmeligi, 2018).

If we peel away the slick surface of our smartphones for a glimpse at what lies beneath, we'll see an array of plastic and rare and precious metals interfacing to produce our machines. Rare and precious metals are harder to extract – the clue is in their name – and demand more inputs than other resources. By their very nature, these metals are tricky. Lithium batteries are the guts beneath our smartphone skin. Cobalt is a key rare and precious metal that forms a compound with lithium to make that battery and it unevenly distributed around the globe.

Let's go to Wilyakali Country, and to a place now known as Broken Hill, to look at a mine that only produces cobalt. In Australia, we produce about 5% of the world's cobalt for predominantly digital consumption. Like so many of the chemicals we use for industrial purposes, cobalt is found in ore bodies of different sorts. Here, the mining company thermally decomposes pyrite orebodies – fool's gold! – so that by-products are minimised, they say.

The so-called Silver City has a relatively long history of mining practices that deepen settler colonial extractive practices in situ and cobalt is yet another that follows these practices. Silver-, lead- and zinc-mining activities precede this wave of cobalt extraction. More than 800 miners have died trying to make a living from mineral extraction in this town, losing their lives participating in practices that dig into, and tear up, patches of Wilyakali Country, land that always was and always will be Aboriginal land.

Walk around old mining sites on the edge of Broken Hill and you can see where these cobalt mines might be heading; the town itself unfurls around a pile of mine tailings. If you climb that mullock to catch first morning light, the surreal notion of building a regional city on mine waste is writ large. The landscape has literally been remade and restructured via mining.

Back to our device for a moment. Smartphones have a short lifespan thanks to built-in obsolescence and the accidents that befall them, such as colliding with asphalt or dunking in water. Big tech makes these devices with a view to rendering them useless before too long. Designing devices in this way means that the high amount of inputs required for smartphones are needed again, and yet again, exacerbated by the limited scope that corporations offer for digital repair and restoration. Recursion really is a drag, to paraphrase Donna Haraway and her Anthropocene musing (Haraway, 2015).

That's just one metallic piece of the digital world we're encountering. Come back to my nature selfie that sits on my phone before I post it to Instagram. Someone chooses to scroll down to my nature selfie, or else they unconsciously participate in procrastinatory infinite scrolling. Either way, that image gets pulled from data centres that store the binary code composing that image.

These data centres might be in the middle of the desert or in busy cities, disguised as factories or semi-industrial buildings. The one we're landing in today is Gadigal Country, close to Sydney Harbour in Ultimo, the largest metropolis in Australia, bejewelled by deep blue water and riven by stark inequities. The Hawkesbury-Nepean, Georges, Parramatta and Cooks Rivers co-produce this city, winding through the sandstone basin to the sea. These physical geographies are easy to overlook as concrete channels and refigured landscapes now usher riparian flows oceanward.

The façade of this data centre is like any office building of downtown Sydney; glass and metal combine to make a commercial frontispiece. Huge amounts of energy are used to run the mega-computers that store the data inside, while water cools the systems down so that overheating does not imperil the machines to keep on storing, retrieving, sharing and forwarding onwards. Fresh water does this cooling – sourced from the aforementioned catchments that masquerade as multipurpose water sources for humans and non-humans.

Wafts of burnt toast intermingle with the ozone from the air conditioner inside these stacks of networked computers. Tubes wrap around grids and the primary colour palate make this data centre feel like a Play School[1] set, but without the joyful warmth of Big Bear, Jemima and the hearty hosts singing nursery rhymes.

In a way, we're lucky that this data centre does not quite approach Google's data centres that are of a scale demanding transport for workers to navigate their insides. If we were in a Google data centre, we'd need bikes to move around their huge structures, if by some miracle we were allowed in. But here, we can meander around the rows of byte-nurturing devices without pedalled assistance.

The data recovery and backup that we rely upon to safeguard our digital lives happens in this Sydney CBD data centre, but not in isolation. We need to travel to data centres further afield, in other countries, and in regional areas, to connect to and understand how this data centre is part of a much larger network of data centres that also consume energy and require water for cooling. After all, no data centre is an island.

The carbon emissions that data centres produce forms a key part of the nature of this digital thing. We won't be able to see these emissions on our visit to the data centre, of course, but we can gauge their scope in, ironically, more data.

Our guide for data centres is Professor John Naughton (2017) who estimates that data centres make up about 50% of all energy consumed by digital ecosystems, or more than 2% of the world's electricity. Another way to think about how much carbon is generated by data centres is to understand that they generate the same amount of carbon emissions as global aviation (in pre-COVID-19 pandemic times). With ever increasing data generation accompanying the intrusion of digital technologies into everyday objects, this portion is likely to increase.

Digital solutions are often put forward for environmental dilemmas – from replacing books with bytes, to enabling citizen scientists to track (and hopefully protect) critters in urban environments. But we can overlook the embedded environmental impacts of our devices that are an important component of our digital natures. Frequently, even basic cost-benefit analysis of digital solutions elude thanks to our modernist inclinations to think that quick, fast, binary data is simply better for everyone, including non-humans.

The nature of digital things is worrying on this front. Data centres are so crucial to keeping digital infrastructure whirring but we tend to think of digital things as more environmentally friendly than using paper or wrangling processes manually. The slippage between 'there's an app for that' and 'that app is good for the environment' is all too frequent.

Data centres could reduce their environmental impact by using solely renewable energy and recycling water for their cooling systems. Apple and Google claim to use 100% renewable energy but this is often through offsets and, while that's a start, it isn't sufficient. Carbon offsets are the sort of climate solution that wealthy people quite enjoy: we can keep polluting the atmosphere with our carbon emissions and absolve our guilt for this continued impact by paying a small price for the offset.

The nature selfie is pulled out again, moving along the tubes and cable that carry it to its next destination.

These tubes often follow pre-existing paths of communication and travel. Even though the digital is novel in terms of the arc of human history, it follows older patterns etched by the so called moderns. Data takes up well-worn paths of unfettered resource extraction and falls into familiar redistributions of the costs of cheap convenience. The routes that cables take, joining data centre to data centre, are layered upon already-present lines of travel and movement.

The Internet is a series of machines linked by wires and cables, supported by satellites and towers. The tubes that carry wires are a crucial part of this infrastructure. Our guide here is Shannon Mattern (2013) who persuasively argues that the wireless Internet is anything but given its hardware. This digital walk is itself inspired by her pedagogical stroll through Manhattan with university students and Andrew Blum, author of a book that unearths the tubes that ground the Internet.

We're travelling in our minds while Mattern, Blum and students took a physical journey to unpack how Wi-Fi actually exists in a hyper urban setting. Mattern and her students went to rooms where networks join together and move apart, and looked at underfoot openings to the subterranean that house tubes aplenty.

We've got our imaginations to digital walk with here. Let's hold up one of these fibre optic cables to the light for a moment. This cable is a part of the circuitry of our digital walk as it enables data to flow back and forth along it, bridging static components of the infrastructure that house and communicate the nature selfie we've been tracking.

The fibre optic cable lets information move around at phenomenal speeds, across geographies that are surprising and complicated. A fibre optic cable is made up of super thin strands of plastic or glass that we call optical fibres. These cables allow for light pulses to push data – our nature selfie – to where it needs to go. They enable data to travel faster and more easily than electric wiring does.

So that exponential data production and consumption that characterises our digital worlds is possible, we've bundled optic fibres together to replace and supplement older technology such as the copper ADSL that we still use in many Australian Internet networks. Remember that debate about whether we should have fibre to the node or fibre to the home when working out what form of faster Internet we'd collectively facilitate? Fibre to the node relies on the node home copper connections.

Weather conditions affect whether that nature selfie will make it to where it needs to go as humidity and rain absorb some of the wireless signals pushing that image onwards. And if my friend is at home, scrolling through their feed and happen to see the small digital world I made, they may have slow connection due to flooding of the digital infrastructure if water gets into cables or connections. Weathering of landscapes affects this digital walk in similar ways to more traditional nature walks. Our path is slowed down as the dirt soaks up rain and our view is impeded by clouds and fog. It may still get to my friend, but it could take a while.

Our path has been unlike your last bush walk or that clifftop coastal trek; we've moved across Country, into the outback, and back to a metropolis. We've visited subterranean parts, come back indoors, moved across ocean floors and slid under concrete footpaths. We might have floated skywards if we'd had the time. There was nothing linear or circular about this meander.

The digital is prone to dematerialisation that generates tricky terrain (McLean, 2020). We can forget that the 'cloud', for example, does consume energy and needs data centres for storage because we imagine the digital as nebulous or virtual. Clouds are ephemeral, digital infrastructure less so.

Our bodies and minds are disrupted by digital things all the time. They are entangled in our daily lives in multiple ways – from smartphones to artificial intelligence devices in our homes, to automation of manufacturing processes and some decision-making. We often assume these digital things work for us and get frustrated when they do not. These are familiar aspects of digital things that we live with.

But the nature of digital things is frequently rendered strange. The key components of digital infrastructure that allow for easy connectivity, anytime, at many places, are underground or in space or innocuous buildings. Traversing ocean floors or in the capacious underground: becoming evident when they fail and at most other times, staying quietly out of mind. The real nature of the digital is invariably hidden by its smooth surface until it is revealed at moments of infrastructural glitch, rupture and breakage. Or if you take a digital walk along with a nature selfie for a while, and follow what emerges in these partial glimpses of digital spaces, and think on possible alternative digital futures and be open to thinking of different digital worlds. After all, as Massey (2005: 12) says, 'For the future to be open, space must be open too'.

Note

1 An Australian children's television programme.

References

Belkhir, L., & Elmeligi, A. (2018). Assessing ICT global emissions footprint: Trends to 2040 & recommendations. *Journal of Cleaner Production*, 177, 448–463.

Carrington, D. (2023). Failure of Cop28 on fossil fuel phase-out is "devastating", say scientists. *The Guardian*, Friday 15 December. https://www.theguardian.com/environment/2023/dec/14/failure-cop28-fossil-fuel-phase-out devastating-say-scientists

Del Casino Jr, V. J., House-Peters, L., Crampton, J. W., & Gerhardt, H. (2020). The social life of robots: The politics of algorithms, governance, and sovereignty. *Antipode*, 52(3), 605–618.

Haraway, D. (2015). Anthropocene, capitalocene, plantationocene, chthulucene: Making kin. *Environmental Humanities*, 6(1), 159–165.

Maalsen, S. (2023). Algorithmic epistemologies and methodologies: Algorithmic harm, algorithmic care and situated algorithmic knowledges. *Progress in Human Geography*, 47(2), 197–214.

Mattern, S. (2013). Infrastructural tourism. *Places Journal*, July 2013. Accessed 07 December 2023. https://doi.org/10.22269/130701

Massey, D. (2005). *For Space*. Sage: London.

McLean, J. (2020). *Changing Digital Geographies: Technologies, Environments and People*. Palgrave Macmillan. https://doi.org/10.1007/978-3-030-28307-0

Naughton, J. (2017). The trouble with big data and bitcoin is the huge energy bill. *The Guardian*. https://www.theguardian.com/commentisfree/2017/nov/26/trouble-with-bitcoin-big-data huge-energy-bill

INDEX

Note: **Bold** page numbers refer to tables, *italic* page numbers refer to figures and page numbers followed by 'n' refer to end notes.

Printed in the United States
by Baker & Taylor Publisher Services